IT'S A QUESTION OF SPACE

IT'S A QUESTION OF SPACE

An Ordinary Astronaut's Answers to
Sometimes Extraordinary Questions

CLAYTON C. ANDERSON

Ing
7/18

University of Nebraska Press | Lincoln and London

Library of Congress Cataloging-in-Publication Data
Names: Anderson, Clayton C., 1959, author.
Title: It's a question of space: an ordinary astronaut's answers to sometimes extraordinary questions / Clayton C. Anderson.
Other titles: It is a question of space
Description: Lincoln; London: University of Nebraska Press, [2018] | Audience: Ages 12–17. | Audience: Grades 7 to 8.
    Identifiers: LCCN 2017050548
    ISBN 9781496205087 (pbk.: alk. paper)
    ISBN 9781496207968 (epub)
    ISBN 9781496207975 (mobi)
    ISBN 9781496207982 (web)
Subjects: LCSH: Astronautics—Miscellanea—Juvenile literature. | Astronautics—Vocational guidance—Juvenile literature.
    Classification: LCC TL793 .A51486 2018 | DDC 629.45—dc23 LC record available at https://lccn.loc.gov/2017050548

Set in Chaparral Pro by Mikala R Kolander.

Contents

Acknowledgments

I enjoy writing. I enjoy it very much. But the success of endeavors such as this one cannot be established without the tremendous support of those helping to pull everything together. And "everything" is a significant quantity.

I must first thank the folks at Quora and their worldwide group of followers who provided the feeder questions through the wonder of social media. Without them—and their platform—this book would not be possible.

Second, I would like to once again say thanks to my family. As I have slowly grown as an author, we have grown rapidly as a family. More separated these days, we've reunited over holidays and college football games to banter about prospective book titles while also sharing laughs, eye rolls, and big groans. Susan, Cole, and Sutton—you make me whole, and I love you for it.

A special shout goes to Annie Balliro, her great folks at Uniphi Space Agency, and my longtime friend and legal guru Joe Montalbano. Your willingness to help me sort through various contract verbiage and to provide concrete inputs for success leads me to offer a huge virtual handshake on our growing partnerships.

Just as those who claim (correctly) that spaceflight is the ultimate team sport, I have several folks to acknowledge on the team that brought this book to fruition. Warren Harold (the man with two first names) was my go-to genius in my quest for locating photos of sufficient quality, sometimes risking the wrath of the National Aeronautics and Space Administration's Johnson Space Center management. Editor Vicki Chamlee performed at a near-genius level as my editor. To my friends at the University of Nebraska Press—Rob T., Rob B., Rosemary, Tish, Courtney, Sabrina, and the rest—thank you for taking another chance on this very Ordinary Spaceman. Let's do it again soon, shall we?

Finally, I would like to thank my fans. To those who read my first book and to those who have ever followed me on social media, I am always here for you—whether it be a quick response to a tweet, a Facebook post, a Quora answer, or a brand-new book. You deserve to know what it's like . . . what it's like to be an astronaut, a spouse, a father, and a friend. Godspeed!

Introduction

The internet and its recent accompanying onslaught of the concept known as social media have had a huge positive impact on societies of the world. They have had a similar effect on me. The human ability to assimilate information and communicate, now unhindered by our world's vastness, has rendered our spherical planet nearly flat. Given electricity, a computer, and an internet connection, there are few significant boundaries to learning.

I began participating with social media platforms in 2008. Starting with Twitter, Facebook, and Instagram, I was introduced to the world of Quora through my good friend and NASA engineering colleague and astronaut trainer extraordinaire, Robert Frost.

Quora has become a wonderful place for me to share my knowledge. Not that I know a whole lot, but I have, after all, lived and worked in outer space for 167 days. I have flown in (or on) space shuttles *Atlantis*, *Endeavour*, and *Discovery*; the International Space Station; and the Russian Soyuz TMA-10 capsule.

Approaching nearly a decade of personal Quora participation and as one of those deemed qualified in answering the questions of others (voted a "Top Writer" each year since 2014!), my "content" has grown significantly. Yet it's not so much detailed factual knowledge I'm sharing but rather my experiences, which have been out of this world. And now, in the pages of *It's a Question of Space*, I hope to share some of it with you. Here, in one single place, you may find insight—if not always straight and technical answers—to the questions proffered to me by individuals from around the world.

I offer these personal tidbits in simple form. Never too technical, they are written with honesty, humility, and a little bit of humor and from my own personal perspective. Many of the questions and answers appear in their original form, almost exactly as they first appeared on Quora.

Others have been edited slightly, and some combined, because I wanted to clarify the question's intent, provide more detailed answers for the readers, and follow the book publisher's different style standards.

The answers presented in this book were written in response to questions posed to me (via Quora) starting in the early spring of 2014 continuing through mid-2017. The questions came to me in no particular order. If someone submitted a question and I had something to proffer, I wrote an answer. After over three years of queries, I sort of hit my stride with respect to how I wanted to participate as an answer provider on the website. The questions as they appear in this book are grouped into various chapters that seemed to make sense to me and the publisher and, hopefully, will position the book as a go-to and easy-to-use reference.

It is my hope that in reading the pages of *It's a Question of Space*, you will acquire increased knowledge, smile a time or two, and gain a new-found understanding of what it's *really* like to live and work in an outer space home that orbits our planet Earth once every ninety minutes.

Oh, and don't forget, as my dear friend and former planetarium curator of the University of Nebraska Jack Dunn used to say, Keep lookin' up!

Abbreviations

ACES	advanced crew escape system
AsCan	astronaut candidate
ATV	automated transfer vehicle (European Space Agency)
CAPCOM	capsule communicator (Mission Control)
EMU	extravehicular mobility unit
ESA	European Space Agency
EVA	extravehicular activity (space walk)
EV1	lead spacewalker
EV2	second spacewalker
FE	flight engineer (1 [lead] or 2)
FGB	functional cargo block (in Russian module)
GLA	general luminaire assembly
GS	general schedule
HTV	H-II Transfer Vehicle (from Japanese Space Agency)
IRED	interim resistive exercise device
ISS	International Space Station
JAXA	Japanese Space Agency
JSC	(Lyndon B.) Johnson Space Center
LON	launch on need
MCC	Mission Control Center
MCS	motion control system
MOM	Mars Orbiter Mission
NBL	Neutral Buoyancy Laboratory
NEEMO	NASA Extreme Environment Mission Operations
NOAA	National Oceanic and Atmospheric Administration
SAA	South Atlantic Anomaly
SAFER	simplified aid for EVA rescue

ShREC	shuttle rotating expedition crew member
SLEEP	sleep-wake actigraphy and light exposure during spaceflight
SLS	Space Launch System
SM	service module (Russian)
SPHERES	Synchronized Position Hold, Engage, Reorient, Experimental Satellites
STS	space transportation system (shuttle)
TCCS	trace contaminant control subassembly
TeSS	temporary sleep station
T_{res}	reserve time
USOS	U.S. Operations Segment

IT'S A QUESTION OF SPACE

The Life of an Astronaut

Question: What's it like to be an astronaut?

Answer: Being an astronaut is indeed the greatest job in the universe. It is challenging and rewarding, full of stress and excitement, and a whole lot of fun. I felt extremely proud in representing my country and undertaking such a noble endeavor that put me in situations and environments experienced by few in human history.

I spent a total of 167 days living and working in outer space. I had the opportunity to perform six different space walks while helping to build and maintain the International Space Station (ISS). I also had the privilege of seeing our beautiful home planet every single day from a vista some 225 nautical miles above it. And I got to perform a multitude of science experiments that may one day lead to advances benefiting people around the world.

For more details on my personal life as an astronaut, you might want to check out my first book, *The Ordinary Spaceman: From Boyhood Dreams to Astronaut*. Here's a short section to give you a flavor of what being a brand-new astronaut is like:

Being a novice astronaut, the phone call that ultimately blew me away was the one I received from the PE (Personal Equipment) Shop at nearby Ellington Field. I was told I would be receiving a "holey joe" (NASA lingo for interoffice mail envelope) containing a couple of forms for me to fill out. Once that task was completed, I would then be able to head out to Ellington Field for a meeting, the content of which was not clear.

For me to be journeying to Ellington Field was a major deal. This was a special, almost revered place where United States astronauts challenged the skies in high performance T-38 jets. So recently bestowed with the title of astronaut, I still considered myself an outsider. I was in awe as I headed for Hangar 276.

Question: Do astronauts have stable family lives?

Answer: Remember, this answer is my opinion and my opinion only. My priorities have always been, first, God; second, family; and, third, astronaut/career.

I have a very stable family life, full of love. My wife and I celebrated twenty-five years of marriage in 2017, and we are very, very proud of the two young people we have raised. But those twenty-five (and counting) years have not been free from stresses and struggles during my career as an astronaut. Some of those stories are told in *The Ordinary Spaceman*. But I have never ever been unfaithful to my wife, never.

The divorce rate is high in the astronaut world. When I became an astronaut in 1998, it was said to be about 35 percent, which I think is way too high. While some may argue that is a normal or reasonable rate for the United States, I counter with the statement that "astronauts are

1. The Anderson family captured in a zero-gravity photograph in the ISS lab module, 2007.

supposed to be the 'cream of the crop,' high achievers that do whatever they set their minds to." So, if that's the case, shouldn't we be able to keep our relationships strong and stable—even through tough times?

Question: Has any astronaut ever desired to stay in space instead of returning to Earth after mission accomplishment? What I mean by "stay in space" is "stay working in the space station and not be left in space to die."

Answer: This is a very interesting question for me and something I had to think about.

When the *Discovery* mission space transportation system (STS)-120 launched in late October 2007, I finally knew I was going home. I had spent nearly five months living and working aboard the ISS, but I worked extremely hard *not* to think about returning to Earth. I needed to keep my focus on the mission and not start daydreaming of reuniting with my family for fear something would happen to upset the apple cart and cause me to make a mistake.

My wife and I had discussed this many times. "What-ifs" abound with regard to the negative possibilities a space launch and mission could entail (think of the *Challenger* and *Columbia*). In the latter portion of my five-month stay, we had both agreed we could stick it out a little longer. By that we meant postponing our longed-for reunion for another month or two would be survivable although not necessarily desirable. Anything much longer than that would have presented a hardship for both of us.

You may ask, Why? Why would an astronaut—living the dream, flying in outer space, pushing the boundaries of science, becoming more famous with every single day and every single tweet—want to come back? Why not stay up longer and do more good for our planet and the people of our world? My answer is simple: I had a wife and children I loved. My place was back on Earth, and Twitter didn't exist then. Don't forget that my spouse was running our entire household all by herself and for much longer than the relatively short time I was in space. She had become the focus of our kids and our family's chief operations officer way earlier when I started to travel back and forth to Russia during my three and a half years of training. She assumed every role: mom, dad, nurse, lawn caretaker, chauffeur, repair person, counselor, financier. You name it, she did it. She had sacrificed much, including her own career advancement, so I could live my dream. It was time for me to be home for her and hold up my end of our marriage vows. After all, she had a career too. To me, that's what love is all about.

Please understand, my time in space was wonderful. It was thrilling, it was incredibly demanding and gratifying, and it was a blast! I would love to go back, but on my terms, not NASA's. My life's priorities are not negotiable, and my family is right there at the top.

Question: What is the typical age of American astronauts?

Answer: NASA has all of those statistics available. Check the *Astronaut Fact Book*, which is available on the website www.nasa.gov.

I was selected at the age of thirty-nine. Others, such as Tammy Jernigan, have been selected as young as twenty-five years old. The oldest selected was John Phillips at around age forty-six; his astronaut classmates called him Fossil. If asked to guess off the top of my head, I would say an average age would be around thirty-five years old.

Question: Why were there so many astronaut applications in 2016? There were more than eighteen thousand applicants applying for the 2017 NASA astronaut class, shattering the record of eight thousand in 1978. What factors attributed to this giant boom?

Answer: I think the answer to why there were so many astronaut applications in 2016 is pretty simple: social media.

When I applied and was finally selected in 1998, some twenty-six-hundred-plus folks applied with me. Only twenty-five Americans and seven internationals were selected. That's a success rate of around 1 percent. I have no idea how many applications were submitted in competition with mine when I applied for the first time, way back in 1983, but I'm guessing it was more than a few. Now in the year 2016—with our hyper-connected world—over eighteen thousand applications were submitted. Further, with an expectation that NASA will choose only eight to fourteen new astronauts (they ended up choosing eight), well, that's a rate of less than 0.1 percent! Tough odds, for sure. My unsolicited advice? Prepare for a few more application submittals before your ultimate selection. Let's see, how many times did I apply before I was selected? Hmmm, fifteen!

Realize, of course, that in those eighteen thousand–plus applications will be some easy ones to reject. Eighty-year-old airplane pilots who don't have the résumé of former astronaut and U.S. senator John Glenn, twelve-year-old girls applying as part of a junior high school project, or individuals boasting of their ability to telepathically communicate with aliens living on the planet Ork will hit the circular file faster than orbital velocity.

But as I, the Ordinary Spaceman, have *always* said, What do you have to lose but a little bit of your time and effort? The rewards can be tremendous, and don't ever tell yourself you won't make it. Let the selection committee tell you.

There is one other thing to consider. It has long been my observation and belief that once members of the astronaut selection committee become involved— and when it gets to the point where mostly current astronauts do the selecting— they opt to pick candidates who are just like them. I am speculating here as I was never asked to participate on that selection committee in my fifteen years as an astronaut.

Question: What is the most difficult task for an astronaut to perform in outer space?

Answer: It's hard to say which task is the most difficult. As with most endeavors, the first time you try something is often the hardest. That was my experience with using the toilet. After the first time (successful, I might add!), my confidence and understanding of what to do and how to do it increased tremendously. The same could be said for going on space walks and for doing maintenance inside the ISS using various tools. Anything I did for the first time necessitated that I read and follow the procedures very carefully to minimize any possibilities for mistakes.

Photography was a task that really took some time to learn. Since the Earth is spinning and the station is traveling around the Earth, our relative velocity was about five miles per second. So with the "ancient" photo equipment we had back in 2007, following or tracking a target with the camera lens, while hoping to capture an in-focus shot, was a bit of an art. I got pretty good at it with time.

Question: Are NASA astronauts only paid between $80,000 and $140,000 per year? Considering that they risk their lives, isn't that salary scanty? Do they get any additional benefits?

Answer: Astronaut pay? Heck, yes, it's "scanty." When compared to TV and movie stars, sports figures around the world, CEOs of Fortune 500 companies, and the like, astronauts don't get paid squat, especially when one considers we do risk our lives. Yet most of us really don't care. We're doing what we've dreamed of doing, many of us since we were young children.

When an astronaut is selected, the salary will range between general schedule (GS) levels of GS-11 to GS-14, based upon the federal government's pay scale and each individual's academic achievements and experience. As of this writing, a GS-11 starts at $66,026 per year and a GS-14 can earn up to $144,566 per year. Military astronaut candidates are assigned to the Johnson Space Center (JSC) and remain on active duty status for pay, benefits, leave, and other similar military matters.

My wife and I discuss this topic frequently. Our conversation always returns to the solid work of our U.S. military veterans. If we consider their working conditions, it puts those of astronauts in a different perspective. I cannot speak for my colleagues, but in my personal experience, astronauts and military servicemen and women have similar experiences in some significant ways. We toil in the framework of government service; we are subjected to family separations, the result of long—and sometimes frequent—deployments. We also must perform according to rules and regulations many would find overly restrictive. Yet we soldier on, many for our love of country and the impact we can have on people's lives, including preserving freedom in America and around the world.

But there is a big difference. As astronauts, we don't put ourselves in harm's way with the regularity of those defending our country abroad. I don't recall any incidents where someone was shooting at me or my crewmates or where I was placed in the unenviable position of trying to find and avoid an improvised explosive device. I never had to patrol a foreign land—wearing heavy and hot equipment that made my space suit seem like a pair of overalls—while protecting those who long to live in freedom and not under some aggressively oppressive regime.

Sure, I risked my life when I dived in the Neutral Buoyancy Lab (NBL) during space walk training. Yes, I risked my life when I flew sorties in the T-38 Talon jets for spaceflight readiness training. And, of course, I risked my life by jumping in an explosive-laden rocket to fly into the vacuum of outer space. But I did it infrequently, and all of those things were *fun*! And when it was all over, people (well, maybe not everyone) patted me on the back and told me how brave I was, how wonderfully I performed during the missions, and how cool they thought astronauts are. I never once worried about spending the rest of my life in a wheelchair or fighting through years of rehabilitation without one of my arms or legs. I didn't spend sleepless nights filled with horrible dreams and the effects of post-traumatic stress disorder.

Certainly, it would have been wonderful to be compensated in the likes of a LeBron James or a Warren Buffett, but it was not necessary for me. I was living the dream. Perhaps it would be better for our brave military men and women to receive more than the meager salaries they get now. And when they do return home? Thank them for their service and give them that pat on the back. They're the ones who really earned it.

Question: Have you gone to space?

Answer: Yes, I have. Twice actually. The first time was 152 days' worth, as a crew member of the ISS from June 8, 2007, through November 7, 2007, to be exact. The second trip was for 15 days as a member of the STS-131 crew. We visited and resupplied the ISS. That trip took place from April 5 through April 20, 2010. During these two trips, I had the privileged opportunity to perform six space walks, three on each mission. Quite cool if you ask me. I believe I'm a pretty lucky guy, and your question is pretty broad, don't you think? In any event, there's my answer.

Keep lookin' up. If you had looked up in 2007 or 2010, you just may have seen me waving at you from the shuttles *Atlantis* or *Discovery* or from the ISS.

Question: How was your experience of working with Sunita Williams?

Answer: Working with Astronaut Sunita "Suni" Williams was a real treat. I have nothing but positive memories of the time we shared together, prepping for our respective turns in space.

Of the first members of our 1998 class of astronauts—the Penguins—assigned to fly to the ISS, Suni was designated to fly there immediately before I was, and I was supposed to be her backup. Originally, she was to fly with Expedition 12, and I was to go with Expedition 13. If all went well, I was to follow her into space and take her place. All didn't go well, with many situations changing NASA's launch plans.

We were labeled by NASA and the Astronaut Office as ShRECs. Not just a clever moniker, the acronym stood for **Sh**uttle **R**otating **E**xpedition **C**rew members and symbolized, to an extent, our roles and responsibilities for the mission content. As ShRECs—Suni was ShREC 1, so I was ShREC 2 (good movie titles!)—we would launch on, and return to Earth on, designated space shuttles. It also meant we were tied directly to different ISS chunks of hardware—for example, the solar array wings, or truss segments; modules; and so forth. In this capacity we were trained to work with these specific pieces of hardware and were likely to get the chance to perform some space walks because of our familiarity with them, a very desirable coup for mission specialist astronauts.

Launch delays and crew reassignments led us to finally be booked for our trips to the ISS as members of STS-116 and STS-117, respectively. I would launch and replace Suni on board the ISS, as her 116 launch via space shuttle *Discovery* left her to become a part of Expeditions 14/15; and I, arriving on *Atlantis*, was part of Expeditions 15/16.

Suni and I called each other Brother and Sister. Our training schedules were such that I took all the training she did and vice versa. We would both fly for differing lengths of time as members of Expedition 15 with Russian cosmonauts Fyodor Yurchikin and Oleg Kotov. And whenever Suni got an itch to do something unique, she pushed me hard to do it with her. After all, wasn't I her backup? These escapades included, but were not limited to, an overnight train trip to St. Petersburg, Russia; winter survival training in the Absaroka Mountains of Wyoming; and many other adventures within the confines of Moscow, Russia.

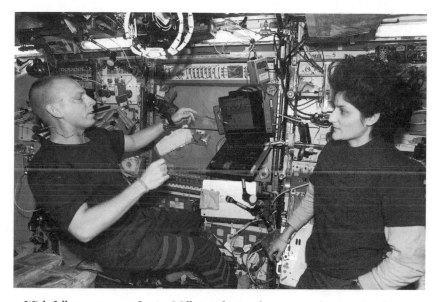

2. With fellow astronaut Sunita Williams during the ISS Expedition 15 handover.

Suni, a former naval officer and helicopter pilot, is a wonderful astronaut. She now holds numerous women's spaceflight records and boasts an extremely impressive résumé. She has a wonderful sense of humor, a passion for NASA outreach, and a slightly happy-go-lucky personality that is reflected in her lifestyle. One of the wonderful privileges afforded me via our training relationship and growing friendship was the opportunity to meet her mother, her father, and her sister. Her mother, my ballroom dance partner at an astronaut function years ago, presented me with a St. Christopher's medal and chain just before my launch. A piece of jewelry that I cherish and still possess, it left my neck only when I was doing my daily ablutions. The necklace, which succumbed to the forces described by Sir Isaac Newton—namely, the severe lack of gravity—on the ISS during one of my towel-bath sessions disappeared at one point and caused me considerable distress. How relieved I was to find it, totally by accident, hovering quietly near the floor behind a cargo transfer bag as I searched for some experiment equipment some two days later!

I would fly on a space mission again with Suni in a heartbeat. She is, and always will be, my "space sister."

Question: If a regular average person were blasted into space with astronauts, would he or she survive?

Answer: Most certainly the person would survive. This is exactly what the Russians have been doing for quite some time now in their Space-flight Participant Program. For a substantial fee, which is way upward of $30 million these days, folks are allowed to fly to the ISS as tourists (although that is probably not the word the Russians prefer). I trained very briefly in Star City, Russia, with Lance Bass, who was one of the first to venture into this program in the early 2000s. While he did not fly, he helped pave the way for others including South African Mark Shuttle-worth and U.S. citizens Dennis Tito and Greg Olsen. As a counterpoint, the United Kingdom's opera star Sarah Brightman was scheduled to go to the ISS in 2015. She reneged, however, causing an unanticipated reshuffle of crew assignments.

All of these participants receive basic training in Russian systems (i.e., how to poop and pee, eat, and sleep in their *kayuta* [sleep station]), Rus-sian language, emergency operations, and survival training. However, it should be noted that Brightman failed to participate in her opportunities for winter and water survival training, making her—at the time—much more dependent on the abilities of her professional Russian cosmonaut crewmates and a bigger liability in the event of a serious emergency. The professional astronauts and cosmonauts will ensure the "tourist" survives; it's part of their job and a huge source of personal pride and responsibility. I would venture to guess all will be well for any and all spaceflight participants heading for space.

Question: What is the astronaut fatality rate?

Answer: Some quote numbers that roughly state the odds of a fatal mission accident for the space shuttle program were somewhere in the neighborhood of 1 in 62. In actuality, during the shuttle program, NASA had two failures in 135 flights. That results in a 1 in 67.5 chance.

During my teaching at Iowa State University, I use the 1 in 62 esti-mate. In the semester's final lecture, I illustrate the odds of death for students using a deck of playing cards. By providing each student with a single card from the deck of fifty-four (including the jokers), the student receiving the ace of spades represents the unfortunate one who doesn't survive. While not exact, it's a powerful representation that gets the student's attention.

Question: How much "me time" do the astronauts in the International Space Station have per day?

Answer: "Me time" for astronauts varies. Depending on your ability to work tasks efficiently, you can create more time for yourself through your work habits. When I lived on the ISS, we had six hours each day to do our specific work tasks. That number was subsequently changed to six and a half hours in the 2010 time frame after some ISS astronauts found it difficult to complete all of their daily work in the allotted time. Unfortunately, increasing that number subtracts from your day's me time as the time must come from somewhere. I found I adapted quickly in orbit and was able to complete most tasks quite early. I had ample time to write journals for my friends and family, to gaze out the windows, to watch TV programs or movies, to call folks on the internet phone, and even to play a Tiger Woods golf video game.

I was also informed during my five-month stay that we were "so busy"—that was the ground control team's opinion, not mine—there would be no time for any space-to-ground media interviews after my first extravehicular activity, or space walk, on July 23, 2007. This news really upset me. I told the Mission Control Center (MCC) that not only did I have plenty of time to do interviews but also this was a special event given my being the only astronaut from Nebraska and taking my first space walk and all. The compromise was that the team on the ground created something called crew discretionary events, which meant that if I chose, I could do these interviews on my own time. Typically they would have been scheduled as part of the six-hour (later the six-and-a-half-hour) workday. Of course, I said yes, and these crew discretionary events continue to be a part of the ISS program today. So I guess they can be credited directly to me.

Question: What type of people live in/on the ISS?

Answer: All types of talented people have spent time living and working on board the ISS. From engineers to chemists to medical doctors to military fighter pilots to helicopter "jockeys," the ISS inhabitants form an extremely diverse group of mostly type A personalities. All are taught the tenets of basic research techniques, and most of the experiments are designed such that those living in space don't have to be true research scientists.

However, with all that being said, I envision one day when the ISS workforce will indeed include research laboratory types and ISS habitation techs (or maintenance personnel) who will be responsible for the day-to-day functions and maintenance of the vehicle's operating systems. Similar to NASA's NEEMO missions to the Aquarius underwater habitat—perched on a coral reef sixty-five feet below the ocean's surface near Key Largo, Florida—a manpower scheme of this type would allow for lab research in the purest sense to be performed, much as it is done in labs all around the world today.

Question: How tough is the interview before getting selected as an astronaut?

Answer: This is a hard question for me to answer. My first interview in 1996 was harder than my second in 1998, after which I was fortunate enough to get selected.

As I recall, the toughest part was walking into the room and seeing all of the faces of the most prominent people in NASA's Johnson Space Center and Astronaut Office lineup. Once past that initial hurdle of celebrity shock (and a quick trip to the restroom to change my underwear), I remember astronaut legend John Young—in his unique and a bit crusty pilot-like voice—saying, "Clay, why don't you tell us about yourself? Start in junior high school and work your way up." Hmmm, the wheels in my brain were turning. What should I tell them? What should I leave out? What do they *really* want to hear?

Turns out I must not have told them what they wanted to hear, as 1996 was my fourteenth year of rejection out of a record fifteen tries. It seems funny to me, but I remember Astronaut Jim Wetherbee, another shuttle astronaut legend, asking a question about the space shuttle and its ascent performance margin as a function of weight, cargo/manifest, and launch window. I couldn't answer his entire question, unfortunately, but I told him I would go back and find the answer he sought. When our paths crossed again a couple of days later but still during interview week, I gave him the full answer, the result of an hour or two of office research. What was his reply, you ask? He said something like, "Oh, that's okay. I really didn't need the answer. I was just trying to act interested." Ugh, that really hurt. But in his defense, imagine sitting through 120 of those interviews, some immediately after the lunch hour and with most lasting around sixty minutes. Good prep for NASA management training.

If there's a key to success in these nerve-racking interviews, I would say you just have to be yourself. Don't put on airs and try to be someone you're not. (Reminds me of the story of the astronaut candidate who refused to sit down for his interview even after being told multiple times it was okay. He didn't fare too well.) NASA wants to see if you are genuine; whether you have "platform skills," or the ability to communicate effectively; and how you respond under pressure. Do you have a sense of humor? Can you clearly and efficiently present your ideas? Can you keep them from falling asleep? Those are the sort of things they are looking for.

The second interview went more smoothly, at least from my perspective. Already knowing what points I would target—based on feedback from my 1996 experience—I focused on a few key factors some "little birdies" told me the interviewers would be looking for. I highlighted my specific space knowledge and skills, focusing on my technical accomplishments from my fifteen years' experience as a NASA aerospace engineer. And I referred them to the annual automobile issue of *Consumer Reports* and its rating of cars for the model year.

To aid those readers seeking to purchase a new car, the magazine's experts use three descriptors: highly recommended, not recommended, and insufficient data. These descriptors were key.

I proposed to the committee that each of them walk the grounds of the JSC at random, stopping any person they encountered. Then I suggested they ask that person if he or she knew or had worked with Clayton Anderson. A majority, I speculated, would respond positively as I had been there a long time.

At that point, I argued, actual data would be available to the committee, just like in the magazine. The members could ask the individuals how they knew me, what kind of person they thought I was, how hard I worked—anything. They could ask if they thought I would make a good astronaut. Would they recommend me, not recommend me, or did they not have enough data to decide?

How'd it turn out? Well, I finally got selected in the 1998 astronaut class!

Question: How difficult is it to get dressed in space?

Answer: Putting on or taking off your clothing—"donning" or "doffing," as NASA likes to say—is a pretty simple task in zero gravity (zero-g). Even climbing into or out of your protective launch and entry space suit is not too bad, unless you are doffing that space suit on the middeck of a space shuttle during your first foray into the extreme environment known as outer space. In that specific scenario, pulling your head through the rubber gasket–lined metal hole that attaches to your space helmet requires a level of physical exertion that can cause the puking fit of a lifetime. Head movement when you are new to zero gravity can lead to being as seasick as during a storm on the ocean. That is not fun at all.

I found that getting dressed in my everyday work clothes was straightforward. The most difficult task for me was putting on socks and shoes. While used to doing this with the assistance of the force known as gravity while sitting down on Earth, floating in a near-weightless environment creates an altogether new perspective on this heretofore simple endeavor. Fortunately, shoes were only necessary when using the treadmill or the weight machine known as the interim resistive exercise device.

With respect to my extravehicular mobility unit (EMU), or spacewalking suit, it was a "can o' corn" (or easy) to don the lower torso assembly. While floating effortlessly in the ISS airlock, it took me only seconds to put my pants on. I remember watching my STS-118 shuttle crew visitors struggling to don their britches while I just floated in the station's crew lock, patiently waiting for them to figure it out. The donning of our high-tech dungarees was apparently an art form learned through extended weightless periods of long-duration spaceflight. The hard upper torso was another matter. To put that shirt on required solid pushing from your legs against the floor, similar to how we do it at the NBL facility on Earth. On the ISS, though, the hard upper torso seemed much easier to climb into. Once tucked inside, I found the suit to be infinitely more comfortable. Rather than experiencing the force of gravity, which due to your weight causes you to sink—and even sit—on the metal rings in the suit, its absence allowed me to float peacefully inside my new "spaceship." Most pressure points that may have been present under Newton's laws of motion—most specifically the one that says when one body exerts a force on a second body, the second body simultaneously exerts a force equal in magnitude and opposite in direction on the first body—then disappeared.

Question: Why do astronauts wear belts in space?

Answer: While this is an interesting question, I'm not sure I have a good answer.

When I lived in space on the ISS for five months in 2007, we did not wear belts. Belts were not offered to us as an article of allowed clothing. I'm sure I probably could have asked for one, but it did not occur to me at all.

You see, back then the primary pair of pants were the royal blues you don't see worn much anymore. They have now progressed to cargo pants; mine were from Cabela's, a Nebraska retailer, and that was really cool for Nebraska's astronaut. The royal blue pants, made specifically for shuttle fliers in the 1980s, had an adjustable Velcro waist, so a belt was not necessary. In fact, the pants had no belt loops, but they did have a custom zippered pouch or pocket on the lower left leg that was a nice place to carry and store needed items during the day. I carried Sharpies and note cards, for example.

The cargo pants of today—shorts are available as well—are still modified with attached Velcro strips on the thighs and lower leg to help us hold pens, pencils, and other small items such as tools. They may even still be the Cabela's brand. The pants have nice wide belt loops, and by using a belt, they do a much nicer job of keeping shirts tucked in. The pockets, standard issue on the cargo pants style, already have Velcro closures, which are important for the microgravity environment. All make for a better presentation of good-looking, well-kept astronauts during those video conferences with Earth. Keep those shirts tucked in, boys and girls.

Question: What's the story behind Clayton C. Anderson's being "kicked to the curb" by NASA's astronaut selection process fourteen times? Did he keep sneaking food into the spaceship simulators?

Answer: The short answer is no. I did not sneak food into the spaceship simulators (although Astronaut John Young did sneak a corned beef sandwich on his Gemini flight in 1965).

But the story about getting "kicked to the curb" fourteen times? Yes. Fortunately or unfortunately, those stories about my personal number of astronaut applications are absolutely true. I applied when initially eligible (maybe even a tad bit earlier than I was eligible) in 1983 and was selected in 1998 as part of the seventeenth class of U.S. astronauts. As I've noted, apparently fifteen times is a record.

I truly believe it took me fifteen years and fifteen applications to be selected as a U.S. astronaut because I lacked the proper "exposure." Simply stated, I think it took me that long to establish my working reputation at the Johnson Space Center in Houston, Texas. I did my jobs to the best of my ability and slowly worked my way into positions of increasing responsibility, with the accompanying added exposure to JSC upper management.

You must also understand that in the astronaut selection process, exposure and credibility may be obtained in many ways. For example, military candidates chosen for JSC visits and interviews often come from short lists that are provided by the various military service organizations—U.S. Air Force, Navy, Marine Corps, Army, and Coast Guard—and go *directly* to NASA and the astronaut selection committee. Then the names on these lists are put before the current military astronauts, who provide a thumbs up or thumbs down along with comments for each name. This process can be very telling.

For civilian candidates selected to come to Houston for an interview—one of *the* key components of the selection process—being the friend, or a friend of a friend, of a current astronaut also can go a long way toward establishing your credibility with the committee.

The key for my no longer being "kicked to the curb by NASA" was my accidentally volunteering to become the manager of the JSC Emergency Operations Center in 1996. In this position—which I initially did not want to take, hence my use of "accidentally"—I performed quite well. I had a blast while doing it (much to my surprise) and gained wonderful, positive exposure to JSC upper management and, as it would turn out, some key members of the 1996 and 1998 astronaut selection committees. The position also got me local TV time, another plus if becoming an astronaut is among your life's goals.

So no sneaking food into simulators was used in the making of this astronaut.

Question: What was it like to work on the "NASA Johnson Style" video?

Answer: I am simply honored to answer this question.

The JSC summer interns (who are now part of the Pathways Program) include both summer and cooperative education students who work for a semester at a time, return to school, and then return to the JSC, hoping for full-time employment somewhere down the road. They invited me to participate in their "Gangnam Style" parody video quite by chance. Someone gave them my name, saying I was "an astronaut who wasn't afraid to do things outside the box," or something similar. I had no idea what I was getting into.

The night before the video shoot, I asked my wife and two kids (then aged seventeen and twelve), "What is up with this 'Gangnam Style' video and who is this dude they call Psy?" My kids were appalled at my ignorance and immediately grabbed the nearby laptop, heading for YouTube to enlighten their aged and obviously pop culture–challenged father.

A musician myself, as I watched Psy's video, I found the tune to be quite catchy, but I had no idea what the video was trying to convey. Questions leaped from my lips to the adolescents in my household. "What's the big deal? What is the point of all this?" I queried. The answers I got back were equally telling: "Dad, there is no point. It's just music and dancing." Go figure.

After having watched the video a few times and still having limited understanding, I reported to the next afternoon's video shoot, scheduled to take place in the bowels of the JSC Building 9. The home of the life-size mock-ups of space stations, robotic arms, space shuttles, and Soyuz capsules, it is one of my favorite hangouts as a full-fledged, training astronaut.

I was greeted by Pathways friends Adam Naids (as I recall, Adam was the original coercer who dropped my name as a candidate for the video); Gary Jordan, essentially the film's director and producer; Eric Sims, the Psy impersonator and star extraordinaire; and others. One of the first questions they asked me was, "Will you dance?" "Of course," was my immediate reply. "Why the heck wouldn't I?"

As these youngsters described what they were trying to achieve and placed me in my starting point of NASA's newest Moon/Mars rover vehicle, I was actually getting a little excited. Truthfully, I was also a bit nervous. Dressed in my blue NASA astronaut flight suit and with my sunglasses perched on my head, I was heading into an audition I wasn't quite sure I was ready for.

With the music blaring so our timing would be right, filming began and fun took center stage. As Eric and I ad-libbed our encounter, complete with my sliding my sunglasses down over my eyes and rubbing my temples to mock Psy's stroke of his jet-black hair in the original video, everything just seemed to flow. It came naturally to me. I just acted like a dork (which my family will readily confirm), watched Eric's moves, and then followed his lead. The true Gangnam steps came too as I had practiced them at home, and I found myself having a grand old time.

We watched the draft video several days later in the JSC cafeteria, and it looked awesome. Synced to perfection by Gary, I felt like a star, if nowhere but in my own mind. Who knew that this video, the brainchild and product of fertile young college minds embedded solidly within the NASA culture, would strike a chord for the ages. It went viral, ending up as *the* most-watched video in the history of NASA—and, yes, that includes all those videos from the Apollo missions, with the guys stepping and working on the Moon—and it became a huge selling point for our nation's space program. Damn, it's good to feel young!

Question: Are the new astronauts who come to the ISS for the first time bullied and teased as Howard was in *The Big Bang Theory*?

Answer: Astronauts tease each other *all* the time. Can you call it bullying? Maybe, but I would say it meets that criteria for bullying only in very isolated instances. As far as teasing goes, I don't know if it's done in the style that ABC TV's *Big Bang Theory* uses with Howard Wolowitz's character, but teasing most certainly occurs.

It starts as simply as giving new astronauts their call sign. Just as Howard was dubbed "Fruit Loops" by real-life astronaut Dr. Mike Massimino (when he accidently learned that Howie's "mommy" was prepping his daily bowl of cereal), rookie astronauts earn their own call signs typically by doing something that is *not* what is called for or expected. Some astronauts in our class of 1998, the Penguins, already had or gave themselves call signs. Those folks included Steve "Swany" Swanson, Tracy "T. C." Caldwell (she later married to become Caldwell-Dyson), and Greg "Taz" Chamitoff. While frowned upon, our class was not worried about ignoring these rules, so we let it slide. Many of the military astronauts already had call signs from their service commitments. As examples, we had Chris "Fergie" Ferguson, Lee "Bru" Archambault, Doug "Wheels" Wheelock, Timothy "T. J." Creamer, George "Zambo" Zamka, and Greg "Ray J" Johnson, among others.

Folks who entered "the game" with no true call sign, such as myself, were dubbed early in our careers—usually by veteran astronauts. I received my first call sign while training in Russia. Astronauts Dr. Leroy Chiao and Bill "Billy Mac" McArthur Jr. dubbed me "Claynus." You see the space relationship obviously. I would later receive a call sign that I enjoyed much better—"Glue." This one came from the fact that my name means glue in Russian.

Astronauts are also teased, and possibly labeled, when they make a mistake or do something really stupid. This can include accidentally driving a T-38 off a runway surface and not coming clean about it. Or perhaps you tossed free pizza coupons to an audience of adoring fans after the owner of the pizza joint gave you "free pizza for life." Assuming you might use one or two of the coupons in any given month, imagine the owner's chagrin when hundreds of people entered his establishment demanding free pizza on the same night.

Astronauts are human beings. We have faults and idiosyncrasies just like everyone else. We work hard while under extreme stress and huge expectations. We have family issues, ranging from infidelity to monetary to medical. It's a tougher job than we let on, and we usually make it look pretty darned easy. Yet, in my opinion, that does not excuse us from accepting the responsibility we have to be role models every single, solitary day.

But teasing? It can be a whole lot of fun if you keep your sense of humor and don't take it personally, but that's another big task for astronauts with type A personalities.

Question: If I were going to visit the ISS, what would be good gifts to bring for the astronauts? I'm thinking food, maybe fresh fruit. Also I'm told in free fall, peoples' nasal passages are clogged, so they like spicy things. Would spicy pepperoni be a good choice? How about fresh cookies, the soft kind, as I know crumbs can be a bother in free fall? Are there things they'd like to have, but NASA and Roscosmos State Corporation for Space Activities won't let them have—maybe chewing tobacco? (I don't think smoking is a good idea.)

Answer: What are good gifts to bring astronauts on the ISS? Darned near anything!

As a long-duration flier in 2007, then having logged 152 days in space, care packages from home were a big deal. What most people don't know is that getting goodies from Earth took a much lower priority when compared with much needed supplies such as food, clothing, spare parts, and ISS equipment needs. And the process to get them loaded, at that time anyway, was complex and involved much negotiation with the Russians, who owned the main delivery ship.

In 2007 our supplies came from two main sources—the Russian's Progress cargo ship and U.S. space shuttles. While the Russian Progress is still the delivery workhorse—launching frequently on a near-quarterly basis—the Orbital ATK *Cygnus*, the SpaceX *Dragon*, the Automated Transfer Vehicle of the European Space Agency, and the H-II Transfer Vehicle of the Japanese Space Agency round out the complement of spacecraft that have been capable of delivering supplies to orbit. The latter two vehicles are nearing the end of their promised agreements with NASA and Russia, signifying the need for Orbital ATK and SpaceX to continue assuming the workhorse role.

Crews are allotted very specific amounts (translated as "not much") for care package content. When I received something from home, I was elated, as it usually was a surprise, similar to opening packages on Christmas morning. Of course, the Russians were pretty good at stuffing a bag of fresh fruits and vegetables into the top of their Progress cargo ship, but I wasn't much for eating apples and tomatoes that looked as though someone had gotten to them with a ball-peen hammer.

I looked forward to photos and letters from home or maybe a gift symbolizing some aspect of my family life. For example, my wife sent me a small stuffed penguin—my astronaut class's mascot—that my mother, who was battling lung cancer, had given to her. Other gifts included

3. Sailing through the ISS with a U.S. food container, stuffed with some family-selected goodies from Earth!

books, magazines, and unique clothing (e.g., a 2007 Houston Astros Craig Biggio T-shirt, commemorating his three thousandth major league hit). The catch was that these items had to be small and lightweight. Carrying large items for crew enjoyment and morale wasn't possible. We were assumed to be steely-eyed astronauts who didn't suffer from psychological issues, so business and mission needs always came first.

For future human missions that are truly of long duration and may involve distinct communication issues—for example, when heading to Mars or beyond—care packages of this sort could be even more important. Yet the method or process for a resupply of this type will probably be much different and involve longer flight times than the quick two-day transit that the vehicles make today. It will be very interesting to watch how this unfolds.

Question: Has NASA ever deemed an astronaut's performance on a space mission unsatisfactory?

Answer: Why, yes, they have, and it was mine! As a member of the fifteenth expedition to the ISS, I apparently became classified as a bit of a "problem child." It had nothing to do with my ability to do all jobs in space correctly, quickly, and professionally. I would learn it was my attitude that was at issue. Many folks on the MCC team felt that while I was doing everything asked of me, I was still not performing as they would have liked. In other words, they were not happy with my willingness to call them out at times, pointing out instances that I thought were ripe with inefficiencies. Their expectation was that I would do everything they said—without question—and keep my big mouth shut.

I cite several relevant examples in *The Ordinary Spaceman*. A short excerpt is provided here:

> Safely back on the ground after a sometimes combative five months, I was sentenced to what I would call the astronaut version of "community service," otherwise known as the astronaut penalty box.
> The words used by the Astronaut Evaluation Board to describe my 152 days of service onboard the ISS were, in part: "Although Clayton is thoughtful with his peers, he needs to improve his communication

THE LIFE OF AN ASTRONAUT 23

skills and attitude towards other teams with which he interfaces. . . .
He tended to be a bit too casual with Mission Control, and sometimes
too frank, and he could have been more patient during stressful times."
They went on to say that "Clay will need to rebuild his relationship
with Mission Control if he is to fly again." The recommendation for
my flight status, as developed by my office peers, was listed as "con-
ditionally eligible."

It's tough to admit, but for some of this they were right. While
my intentions were always aimed at making things better for those
who would follow me into space, I had not heeded the advice I'd been
given and I let the frustration build to a point where it affected my
work and my interactions with the ground.

Yet I wasn't totally at fault. The situation on ISS where we were all
assigned work behind the same panel in the same week was ridiculous.
As a crew support astronaut for the Expedition 4 crew, I participated
in the weekly planning meetings where these types of situations were
discussed. On numerous occasions I was the "elephant in the room"
who complained when the technical team failed even then to grasp the
concept of "proper planning prevents poor performance." To direct a
crew to waste that amount of identical (and expensive) crew time on
orbit was the highest form of government waste. It was inexcusable.

Even though my family and I had some legitimate grievances, I
could have handled myself better. I did not follow the unspoken rule
that no matter what, the ground is always right and they should be
treated with kid gloves.

I took to heart the "community service" recommended by the Astronaut Evaluation Board "that Clay would benefit from leadership/
followership and teamwork training," that "he be put in a leadership
role, perhaps as a branch chief, to satisfy this development in part,"
and that "he consult with his Human Resources representative for
additional development classes."

Still, it was tough getting dressed down like that.

I was not taken off probationary status until I received a call from
astronaut management in "the corner office," telling me I had been
assigned to the crew of STS-131 and *Discovery*.

Question: Do astronauts need passports when they travel to space and leave Earth? Shouldn't they have them, at least when their mission could imply their having to land on Earth in any territory that isn't their own?

Answer: Astronauts do *not* need passports or visas when they leave Earth and travel into outer space; at least they don't at this time. Astronauts launching from and landing in the Russian Federation *do* need their passports when on the ground there. Prior to launch they are typically given to administrative personnel, who ensure everything is in order for the crew members.

On a slightly related note, a good U.S. astronaut story concerns first-time-flier astronauts heading toward the launch pad for their inaugural trip into space. As a way to indoctrinate these so-called rookies—so the story goes—experienced astros would cleverly have fake shuttle boarding passes made in advance. All veteran fliers on the crew would conceal their boarding passes in one of the numerous pockets adorning their bright orange launch and entry suits. Then on their way to the pad and just before the final drive up the hill to the launch crawler platform, the commander sternly would instruct the crew to "have your boarding passes ready." Imagine the confusion of the rookies as they scrambled to try and figure it out and asked, "What boarding passes?" The looks between the already stress-laden neophytes are said to be priceless.

With the trap sprung, good laughs resulted as the commander gave up the sham. Hopefully it provided a bit of stress relief for those heading toward their initial meeting with microgravity.

Unfortunately (or maybe fortunately?), the commanders for my two flights into space—Rick "C. J." Sturckow (STS-117) and Alan "Dex" Poindexter (STS-131)—chose *not* to utilize this worn-out shuttle tradition. It might have helped me calm down!

Question: Are the space suits that astronauts wear for publicity pictures real? From what I've read, a space suit weighs about 110 pounds, and when you add on the full life support system, it jumps to more than 400 pounds. How do they manage that for Earth-based publicity pictures?

Answer: When astronauts and, for that matter, cosmonauts pose for their official portraits, they are wearing actual space suits. Much care is taken during these photo sessions, with the doffing and donning of the suits being supervised by suit technicians. They are also there on launch and landing day to help with the white Russian Sokol or the orange U.S. astronaut's advanced crew escape system suit. Suit techs also support all space walk training, which requires the white EMU, for our U.S. practice runs in the NBL's swimming pool and the Orlan suit runs in Star City, Russia.

When taking photos of astronauts in the EMUs, two technicians are actually hidden in the picture, helping to prop up the incredibly heavy suits for the photo session. They go above and beyond (below the suit anyway) in the call of duty.

Question: When writing memoirs, how do authors remember specific details of past events and scenes? There are often *very* detailed descriptions of minutiae. Is this all pretty accurate, or is some or most fabricated for the sake of setting a scene?

Answer: When writing my first book, *The Ordinary Spaceman*, I used a combination of memory, research, and shorter journal entries I had kept during my fifteen-year career as an astronaut. In some specific instances—such as a meeting with management where I got dressed down a bit—afterward I immediately wrote down notes on exactly what happened. That way I had the information available when I covered those incidents in the book.

For many of the stories, I referred to training documents and mission-specific publications—for example, spacewalking time lines—I had retained from my astronaut career. I even contacted friends in the Johnson Space Center's Public Affairs Office and Astronaut Selection Office to secure appropriate and updated information. It was very helpful to refresh my memory through reviewing various documents, thus enabling me to provide as much credibility as possible to all of the tales.

I am not sure just how detailed one might consider the stories in my memoir, but I think they are of a reasonable and enjoyable level. I did *not* fabricate anything for the sake of setting a scene. Yet, there will be some, I'm sure—perhaps even those mentioned in the pages of *The Ordinary Spaceman*—who will undoubtedly mutter, "It didn't happen like that." To those questioning few, I would challenge them to step up and provide their version, as I honestly endeavored to recollect everything to the best of my ability.

I hope you all will read the result. It's a fun and easy read.

Question: Can astronauts cast their votes from the International Space Station?

Answer: Astronauts most certainly can cast their vote from the ISS. As a matter of fact, I did just that in 2007. And making voting from space even cooler (at least to me) is that my beautiful bride, Susan H. Anderson, was *the* person who made it happen—way back in the shuttle/Mir days.

Astronaut David Wolf—an Indiana native but a Texas resident at the time—was the first American astronaut to cast his vote from space. A resident on the Russian Mir space station during the voting cycle, his ability to cast his vote from space rode on the pen of then Texas governor George W. Bush. My wife led a NASA team of individuals who developed the protocols needed to ensure the space flier's voting process maintained appropriate privacy and allowed orbiting U.S. citizens an easy way to cast their ballots. Files are encrypted and the process tightly controlled, minimizing the number of "earthlings" that need to be involved.

In 2007 when I exercised my voting rights as a citizen—a bit removed from my local polling venue—the vote I cast was for local ballot items, and there were no contests for higher office.

As we look forward to those times when many Americans—and perhaps citizens of other spacefaring countries—will be analyzing their own personal choices to vote in local, state, and national elections, let us remember that for Americans, voting is a right we are granted by our Constitution. Make your vote count, every single time. Not registered to vote? What the heck are you waiting for?

Question: What is it like to be in a shuttle during liftoff?

Answer: Following you will find an excerpt from chapter 12 of *The Ordinary Spaceman*. Enjoy, and, no, I didn't include the entire chapter. Buy the darned book, please!

Through the hard work of thousands of people, we were ready, holding at nine minutes before launch, the final time to decide whether the shuttle is indeed ready to go.

The person making that decision would be the launch director, Mike Leinbach.

My attention was now solidly on the communications pulsing through my headset. Listening to every word, without a hint of needing to pee, I heard Payne initiate the launch status check. Polling his team of controllers, he asked for their readiness to resume the count and whether they were "go" for launch. Unanimous calls of "go" preceded Payne notifying the launch director that his team was indeed ready to proceed.

With that, Leinbach conducted his own poll, pinging the engineering, safety, weather, payload, and operations management teams. Hearing nothing that would preclude an on-time lift-off, he informed Payne that we were "go for launch." The countdown resumed, with the test director's call, "In three, two, one, mark!"

With only minutes until liftoff, I opened my small Ziploc bag that contained my antinausea meds. I popped the pills and took a swig of water from the bag precariously strapped to my right knee. They would enter my system in ample time to aid my first interactions with zero-g.

The crew was silent. Commander and pilot responded only per the checklist in the performance of switch throws and calls to the launch team. Essential electrical circuits were connected to the electricity-generating fuel cells. Auxiliary power units were started, giving life to the vehicle's aero-surfaces being counted on to help steer this powerful stack of rocketry. The caution and warning database was cleared of all previous errors. We had a clean slate. Focus was on the coming liftoff.

4. The space shuttle *Atlantis* launches from Kennedy Space Center, June 8, 2007.

I wiggled and squirmed to find comfort. I was feeling like a multilayered deli sandwich on flatbread. Hard on my back for over two hours now, on layers of parachute, harness, spacesuit, and bladder, stacked on an aluminum-backed chair with a puny half-inch pad, I was ready to be free.

It wouldn't be long now.

Listening to the constant chatter of the launch director and his team, I drew a mental picture of the passing milestones leading to ignition. My

5. Preparing for launch by donning the advanced crew escape system suit in the suit room on launch pad 39A, 2010.

excitement continued to build as the shuttle's onboard computers took control of *Atlantis* and the clock continued its downward trek to 00:00.

At T (time of ignition) minus two minutes, the orbiter test conductor called: "Close and lock your visors. Initiate O_2 flow." We replied collectively, through the voice of our commander. "*Atlantis*, Roger," was all C.J. needed to say. Now that my visor was locked, the breathing sounds in my headset reminded me of Darth Vader. It was time. We were going to space!

Hours of simulations had me ready for the cadence of the flight deck crew. With the commander's first call of "Nav Init," we had less than ten seconds. I prepared to become a spaceman.

My body tensed with the roar of the shuttle's main engines rapidly coming to power. The vibrations from these mighty nozzles coming to life in the six seconds prior to liftoff shook the entire shuttle and filled our comm-cap-covered ears with thunder.

"Engine Start . . . three at a hundred." Pilot Lee "Bru" Archambault confirmed what the shaking and noise had already told me—our engines were purring at 100 percent.

"One-oh-two, one-oh-two, auto, auto," came the simultaneous calls from Commander Sturckow and flight engineer Steve "Swany" Swanson, matched exactly with a countdown clock of 00:00. The shuttle flight software was correctly doing its thing, and the solid rocket motors were fully ablaze. We were going somewhere!

As C.J. called, "Clear of the tower," Bru chimed in with, "Three at one-oh-four." We were traveling 120 miles an hour, the shuttle's main engines working together at 104 percent capability, essentially full power and then some, cranking out 394,000 pounds of thrust each!

In less than a minute, we were accelerating faster than the speed of sound and burning propellants at the rate of ten tons per second in our highly controlled explosion.

The call of "good digitals" came solidly from Swanson, indicating the primary flight control software was seeing the same thing as the ample but less capable backup lines of code.

C.J. contacted Houston. His voice vibrating, akin to talking while riding in a helicopter, he offered, "Houston, *Atlantis*. Roll Program."

Houston acknowledged the computer-driven roll of the multi-element stack to a "heads down" attitude, meaning that the antennas mounted on the orbiter's roof could now contact ground stations with stronger signals. "Roger roll, *Atlantis*," CAPCOM [capsule communicator at Mission Control] and C.J.'s fellow marine Tony Antonelli replied.

"LVLH on the left."

"LVLH on the right."

The sequential calls from commander and pilot signified completed switch-throws on the cockpit's front panels. Shuttle displays would now show attitude data in a local-vertical, local-horizontal reference frame, which is much easier for us earthlings to comprehend.

It was the greatest roller-coaster ride in the history of man. We rumbled, we shook, and the sound was overwhelming. Anything not solidly tied down fell to the floor as high fives and fist bumps passed through the middeck. It *was* the experience of a lifetime!

2

The Basics of Outer Space

Question: Can water and other drinks that "run," or float, away while astronauts on board the International Space Station try to consume them damage the electronic equipment on the ISS?

Answer: If you are referring to what happens when astronauts "play" with liquids while living in a near-zero gravitational environment, yes, it is possible for those free-floating liquids to get into sensitive equipment and cause significant damage.

While I lived on board the ISS, one of my biggest concerns was about shuttle crew members who arrived in space with little to no experience in microgravity. Most *loved* to play with liquids, yet they didn't really

6. The physics of surface tension displayed on the shuttle *Discovery* middeck.

have the experience to pull it off without making a mess or doing damage. The other issue was their desire to always use tropical fruit punch as its bright red color made for the best pictures and videos. Well, spilling red punch all over everything leaves many stains that can be a real pain to clean up.

Quick reflexes are often required as the slightest force on the bag of liquid or the straw or even the floating bubble could cause a liquid disaster. My favorite trick was to create a large sphere of liquid, let it float slowly in the middle of a module, and then gently blow on the sphere to move it where I wanted it to go. We could also put things inside the sphere, such as M&Ms or Japanese flower blossoms. That always led to some spectacular video clips and photos.

Question: What do you think of people who claim that the Moon landings were a hoax?

Answer: Hmmm, I guess a bunch of folks think that all of the Moon landings were hoaxes. Maybe they are also part of the group that thinks Earth is flat as well? What do I think of folks like that, you ask?

Well, in today's world I should probably be ultracareful about what I say regarding folks who think those sorts of things. But then again, since I'm known as the Ordinary Spaceman—after the title of my first book—I typically say what I think (sometimes to my detriment). So my response to them? Tighten up your tinfoil hats (or your athletic sneakers, as the case may be), put on your "big-person pants," do some research, and pull your heads out!

Oh, yeah, and once you have your heads out of Uranus (see how I worked in that space reference?), keep lookin' up!

Question: How many rotations does the ISS make per day around Earth? Does the number change?

Answer: The ISS flies around Earth in a near-circular orbit. Today, the current altitude for that orbit is somewhere around 250 nautical miles. Given that altitude, the period, or time for one orbit, is approximately ninety minutes. With a ninety-minute orbit and a twenty-four-hour day, the ISS will circle Earth sixteen times a day. That's sixteen sunrises and sixteen sunsets each orbital day, and it makes for some great picture-taking opportunities. On most days, however, the crew is far too busy to see all of them.

Question: What's it like to see sunrises and sunsets every hour and a half when you're in the International Space Station?

Answer: Honestly, for me it actually got old after a while. In truth, we are usually working so much that we don't see many of them. If we had a break in our schedule and could look out one of the side-facing windows (we didn't have the cupola during my time on orbit in 2007), it was fun to watch a sunrise or sunset and try to capture the beauty and fragility of Earth's thin atmosphere as the sun rose or set along the horizon. Sleeping at night in our kayutas, we (the U.S. segment) have no windows, so the continual rising and setting of the sun has no effect on our sleep. The kayutas in the Russian part of the station *do* have a single window in each and, smartly, metal covers to rotate over them at night.

Question: How much does an average person understand about space and space exploration? It seems that not many people understand what is out in space and what it means to go on space exploration missions. Does anyone know—for example, through statistical information—how much the general population knows and understands about space exploration?

Answer: If the questioner is correct, I am a bit disheartened. However, there are ways for the average person to improve on this understanding. That some are already asking questions and reading answers on the internet—although not all are factually correct—is a great start.

I would recommend that people check out various social media platforms—Quora, Facebook, Twitter, Instagram, Pinterest, and so on—and astronauts online. You can follow Michael Lopez-Alegria, Garrett Reisman, Doug Wheelock, T. J. Creamer, Richard "Rick" Mastracchio, Reid Wiseman, Karen Nyberg, Steve Swanson, Barry "Butch" Wilmore, Dan Tani, Nicole Stott, twins Scott Kelly and Mark Kelly, Chris Hadfield, Leroy Chiao, Ed Lu, Anousheh Ansari, Jeremy Hansen, Alexander Gerst, Luca Parmitano, David Saint-Jacques, Russian cosmonauts, and many others.

In addition, a considerable number of NASA engineers, flight controllers, flight directors, and public affairs experts are also active on social media. They all aim to educate the public about what we do at NASA and why our work is important and a benefit to all humanity.

Question: What happens when satellites or astronauts go through the South Atlantic Anomaly (SAA)?

Answer: It just so happens that we cover this very topic as part of my class Introduction to Aerospace Engineering at Iowa State University.

The SAA is an area where the Earth's magnetic field is considerably weaker. Varying with altitude, this results in a region in space where we are not protected by nature's "built-in" magnetic shield, and more high-energy particles—ions, electrons, and protons—are able to penetrate within, posing potential dangers to satellites, other orbiting spacecraft, and astronauts. From a human's perspective in space, if the trajectory of the ISS shows it is scheduled to pass through the SAA on any given day, the odds are good that several consecutive passes will touch or pass through that SAA footprint. During periods when solar activity is high—or solar maximums, which occur on an eleven-year cycle—astronauts may be directed to take refuge in the Russian segment or the Soyuz capsule, which offers them a slightly higher level of protection than just hanging out in the station.

Equipment can be impacted as well. For example, ISS laptops may experience single event upsets from high-energy particles impacting their circuit boards. Often a simple reboot can fix that problem, but in my time on orbit, I also had to replace numerous computer hard drives.

For additional protection, some astronauts have lined their sleep stations with bags of water, as good old H_2O provides excellent—and readily available—protection from the bombardment of high-energy particles. Others have also buried their personal dosimeters, which measure exposure to ionizing radiation, in between water bags, hoping

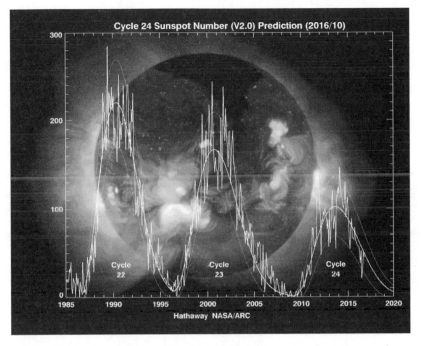

7. A graphical representation of the historical maximums and minimums of our sun's eleven-year solar cycle.

to reduce their overall numbers of radiation dosage and to artificially keep them healthy enough to stay in the program and one day fly again.

Further, the planning of space walks is done very carefully. We make every effort to ensure that astronauts working outside of the ISS are not unnecessarily exposed to the increased radiation of the SAA, as our suits do not provide sufficient protection from these high-velocity particles. Planners must analyze trajectories and solar activity data to determine when the ISS will pass through the SAA and then schedule the outdoor excursions to avoid those calendar days.

This is another good example of how the hundreds, if not thousands, of dedicated people on Earth help those of us privileged to journey from Earth. Spaceflight is the ultimate team sport!

3

ET Phone Home

Answer: What is NASA hiding from the world about aliens? Beats the heck outta me!

I have found many folks are keenly interested in this subject. As the Ordinary Spaceman or "Astro Clay," I have nothing significant or secret to contribute, but I can offer my own thoughts for your consideration.

I am often asked about UFOs. Did I see any? If so, what did they look like? Did they communicate with me? These questions illustrate the first part of my issue with all of this. Folks confuse the term "UFO," or unidentified flying object, with aliens or—perhaps to be clear—extraterrestrials (ETs). That is, someone or something not of our Earth.

I *did* see a UFO during my first space walk on July 23, 2007. I was flying on the end of the Canadian robotic arm (dubbed Canadarm2), which was being skillfully piloted through space by my crewmate and Russian cosmonaut Oleg Kotov. The UFO appeared to my left, silently floating with a very, very slow spin rate. As it spun and the sunlight caught its reflective surface, it flashed brilliantly, as if someone periodically turned on a small light. The object appeared to be about the shape of a large butterfly and, to my eyes, was apparently a piece of metal. I had no idea where it came from. Was it part of the International Space Station? Maybe another spacecraft? Had we dislodged it from somewhere as we were performing our seven-hour-and-forty-one-minute space walk? I had no clue. After all, to me it was an unidentified flying object. I probably should have notified the ground control team about it, but I declined, knowing that its retrograde trajectory, slow speed, and small size would pose no threat to our home in space.

As for aliens or ETs, that's a totally different story. I was never contacted by any, nor did I put my cranium solidly against an ISS or shuttle window and silently try to contact them via telepathy as some of my astronaut colleagues are thought to have done. If they were out there and they were trying to contact me, I would have answered. Trust me on that one!

Our universe is huge; it's too big for our human brains to even imagine. Think of it as a humongous pizza. We, from the planet that first visited the Moon, are on one edge of the pizza's crust. In our 200,000-year history on this planet, we humans have ventured about 238,000 miles from Earth. Consider another species—whose space accomplishments and knowledge rival ours—exists on the exact opposite edge of the crust. To me, it's gonna be an awfully long time until we meet at the pepperoni in the middle! And let's face it: if there's an alien species out there capable of traveling through space and covering distances of light years to Earth, then they're probably intelligent enough to contact us any time they want.

Will "first contact" ever occur? Has it occurred already? I don't know. But if it has, NASA determined long ago that I didn't have a "need to know." As one of the privileged humans to have visited outer space, though, I'm guessing we're still waiting for contact and that, yes, someday it will happen. When it does, I will probably no longer be around. But in the meantime, *you* should keep lookin' up. Someone—or something—may be trying to contact you!

Question: Is it true that almost every astronaut has seen a UFO while traveling to the International Space Station and back?

Answer: No, it is not true that *every* astronaut has had the experience of seeing a UFO while traveling to the ISS and back. But as UFO means an unidentified flying object, I'm guessing you are actually referring to an alien species or an extraterrestrial.

I did not see any of ET's relatives during my two trips, first in 2007 and then again in 2010, to and from the space station.

Honestly, I was open to making first contact with an alien being or inhabitants of a UFO, but it simply didn't happen. As I said before, while

I didn't stick my head against one of the windows of the space station or the shuttle to enhance any telepathic signals coming my way, I did not turn any away either. I was more than willing to "answer" if they called, but, alas, the lines stayed quiet during my 167 days in space.

Question: Are there any rules or procedures issued by NASA or any other space agency for the eventuality of astronauts meeting aliens or UFOs?

Answer: If NASA has developed any rules or procedures for the time when astronauts one day meet aliens or encounter UFOs—and for those of you out there that think space debris is a UFO, I hear you, but I'm talking about an ET—I don't know about them. Perhaps I missed that training class when I was working.

So my answer is simply and unequivocally, no. I do however have some personal procedures I will follow if I am graced with that wonderful opportunity. First, I will smile and hold out my hand to them or it or whatever. If I am still alive, having not yet been engulfed by their massive jaws or sharp teeth or zapped by some advanced space laser, I will speak. Perhaps I will say "hello" or maybe "welcome." I really don't know at this point, but I'll think of something. Assuming all is still going well, I will try and further communicate with them by showing photos or using hand signals. Heck, they might already be able to understand English or Russian.

In the event they are not "friendlies," I will take whatever comes, knowing I have been lucky enough to interact with beings from another world. Don't hold your breath, though; personally, I think it is a long way off.

4

The International Space Station

Question. Why do people live on the International Space Station?

Answer: People don't just live on the International Space Station. They live *and work* on the ISS. Today the home in space I helped build in 2007 and 2010 has become an international laboratory, one whose discoveries may help to change the face of our planet Earth.

My time on the ISS was spent working on an enormous number of tasks—maintenance, construction, science, and otherwise—but at that time, science was not the highest priority. Today it is. Astronauts and cosmonauts from around the world (like how I slipped that reference in?) are working together with scientists on terra firma, hoping to answer a score of questions that could have bearing on how we live and work in the future.

For example, astronauts on the ISS have recently sequenced DNA. We are also studying the effects of microgravity on the optic nerves in our eyes. Techniques are being refined to grow better crystals, and we are evaluating how fire behaves in a near–zero gravity environment. But we've yet to hit the mother lode—that is, *the* single discovery that will help us turn a corner in solving an issue that pervades human physiology or will unlock a secret of our universe. But we're working on it.

I am a huge supporter of NASA, yet I am also a taxpayer who wants to see real and measurable paybacks. I know it's cool that bowling ball-sized spheres can rendezvous and dock with each other inside the ISS. I know it's cool how surface tension allows astronauts to take their "coffee in a bag" and pour it into a cup, allowing them to smell its wonderful aroma and drink it as if they were back home. And I know it's exciting for scientists to gather millions of data points while searching for the black matter believed to be a key to the beginnings of our universe.

I truly believe that if we could find a cure to something, or the answer to a burning question about the origins of our solar system or universe, or anything of consequence, we might never have to worry about our NASA funding again. We were close once, nearly finding a vaccine/cure for salmonella, and we are working hard on perfecting the process of turning urine into fresh drinking water.

So when you ask why people are living on the space station, I challenge you to do it with your eyes toward the heavens and with an adaptable, inquiring mind, one that asks perhaps a different question. Ask, How close—exactly—are we to telling the world of a truly magnificent discovery? I'm hoping NASA can respond quickly and clearly. But I'm guessing, my friends, that may take more time than I'd like.

Question: How is the interior of the International Space Station illuminated? What kinds of lights are used?

Answer: How is the ISS illuminated? With lights.

Actually, in 2016 a combination of solid-state lighting assemblies and general luminaire assemblies (GLAs) are used. The solid-state lighting assemblies are light-emitting diodes, and the GLAs are essentially fluorescent bulbs.

During my time on the ISS, we only had GLAs. They were located in what are called standoffs in the U.S. modules. The modules have standoffs at the both the top and bottom, but the GLAs were mounted at the tops. In this way, as with our homes on Earth, the lights were near the ceiling. It was thought that by doing it this way, astronauts would more readily adapt to microgravity, as seeing lights near the ceiling rather than the floor is more normal.

GLAs are made up of two parts and could be replaced as necessary. The baseplate ballast assembly mounts to the ISS structure and contains the on/off push buttons and dimmer switch. The lamp housing assembly encloses the bulb. I joked often with NASA that we should change the GLA name to the long, incandescent, glowing hot thing (LIGHT), but they didn't think I was very funny.

When changing a GLA in space compared to here on Earth, the major difference was that the actual bulb is enclosed by the housing assembly,

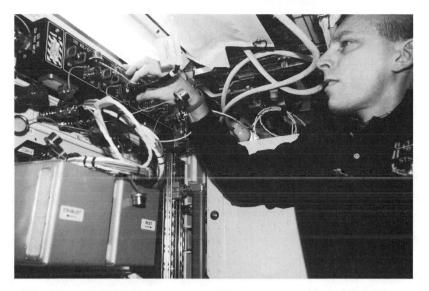

8. Checking out equipment near a general luminaire assembly (light) in the station's U.S. laboratory module.

so broken glass would never be a problem in microgravity. We'll keep the lights on for you!

Question: Is the space station's design for long-term occupancy purely functional, or were there psychological tricks used to make adjustment easier and more livable?

Answer: I can't speak as an expert on this one, only as a resident. The physical design of the ISS is more functional than psychological, but some things were done in an effort to make the space more livable. For example, the Node 1 module Unity is painted in a salmon color, supposedly to make it a more "calming" module, while the U.S. laboratory module Destiny is painted or trimmed in a bright blue color. Mostly though, it's pretty sterile, with white and gray shades dominating the interior. The Russian service module (SM) Zvezda is even less inviting in my opinion. Its hues are more brown, pale green, and tan, all quite subdued.

Lighting is essentially similar to that of rooms on Earth. Studies and designs are in the works to make the lighting more controllable than on/off, thus allowing the crews to shift their circadian rhythms more effectively by varying the lighting intensity with time. The interior sounds are basically those of moving air, operating fans, and pumps. This provides quite a high level of ambient white noise on a constant basis. If it gets quiet, the crew needs to worry as something is probably broken!

Music and movies are the entertainment choice of the many space fliers. We were allowed to have our favorite tunes loaded onto CDs for the trip into space. I loaded very few; other astronauts loaded thousands of songs! Movies were sent in a huge DVD pouch with hundreds of shows. We were able to bring a few pictures, and we received care packages from the various ground control teams, which can also send things for us to customize our tiny living quarters.

All in all, I found the ISS quite cozy and a great place to live and work.

Question: Is there a melody that inspired you during your stay in the International Space Station?

Answer: Actually, a couple significant melodies inspired me during the 152 days I spent living and working aboard the ISS, but one stood out from all the rest.

I stumbled upon my favorite by accident. Not having sent a ton of my own favorite tunes to be preloaded onto CDs for use in space, I resorted to using the files left by my predecessors Mike "LA" Lopez-Alegria and Sunita "Suni" Williams. Their hundreds of tunes were more than sufficient for me to perform my own version of daily micro-g karaoke. I don't remember what day it was or how it happened, but very early in my increment in 2007, I would listen to—and enjoy immensely—a song by the American punk rock group Green Day. The song called "Good Riddance (The Time of Your Life)" is from the group's fifth album titled *Nimrod*. The song, via that album, was first released ten years earlier in 1997.

My musical tastes, as well as those of LA and Suni, must be pretty good since the song went on to sell over 2.6 million copies!

I found the musical tune addicting, as well as uplifting, and I especially related to the words of the chorus. Living out my lifelong dream of being an astronaut, floating through the ISS as if I were Superman, I would play the song often. Singing along, especially on days when I might be having a rough time, it always served as the pick-me-up I needed to get through the remainder of the day.

After all, it was "something unpredictable." I was sailing some 225 nautical miles above the Earth, living with two Russian cosmonauts, and away from everything I knew.

And in the end? It "was right," and I certainly did have the time of my life!

Question: What is the typical temperature range in a spaceship or space station? Do astronauts control the internal temperature on the ISS just as we do in our homes on Earth? Or is the thermostat controlled by the Mission Control Center?

Answer: I am asked this question a lot. My answer, while being a tad bit flippant, is still pretty accurate: While living on the ISS we typically experienced an environment that consisted of a temperature of about 72 degrees Fahrenheit (22 degrees Celsius), humidity of 55 percent, a cloudless sky, and a slight breeze from the south!

Typically, the ISS temperature in the U.S. Operations Segment (USOS) is controlled by the MCC team. The flight controller for the environmental and thermal operating systems has the capability—with concurrence from the ISS crew and the ground team, including the flight director—to set the internal temperature. The temperature of the Russian segment is controlled by the ground control team in Moscow.

The environment I described was maintained during my time on ISS Expedition 15 with my Russian crewmates, station commander Fyodor Yurchikin and Soyuz commander and station flight engineer 1 Oleg Kotov. When the new commander Peggy Whitson and her Soyuz driver Yuri Malenchenko arrived, Peggy wanted it to be warmer—which felt like quite a bit warmer to me—but I am not sure what temperature she requested exactly. It was higher than 72 degrees, for sure. Astronauts who want to change the temperature level can do it through a space station computer's environmental control display or by simply asking the ground control team. Calling the MCC is the preferred method. During my time in space in 2007, I never messed with the internal temperature, nor did I want to.

The "slight breeze from the south" was due to the constant air movement caused by the interworking of the environmental control systems in the U.S. and Russian segments. Combined, they "pushed" the air from the back of the ISS, or the Russian SM; through the "plumbing" to the front of the ISS, or the U.S. lab module; and then back down the center of the modules to start the process over again.

Humidity seemed to be comfortable most of the time in the U.S. segment, but all modules were not the same, with it seeming to be higher in the Russian segment. The coldest module was the U.S. lab, followed by the Russian SM. The Russian functional cargo block was the warmest, with the U.S. Node 1 being the next warmest, in my opinion. Modules off the center axis—for example, the U.S. airlock Quest and the Russian docking compartment-1 Pirs—seemed to be nice and cool most of the time. As my father used to say, "If you're cold, put on a jacket."

Question: If an astronaut murdered all of his or her companions up in space, what would happen?

Answer: If an astronaut murdered everyone up in space, he or she would be pretty damned lonely for a while, that's for sure!

As ludicrous as your question is, my 167 days living and working in outer space allow me some perspective with which I offer a tiny bit of insight into the scenario. Given the level that astronauts are prescreened for maladies—both physical and psychological—and my intimate knowledge of the astronaut corps, I can confidently say it will never, ever happen. But if it did, I'm guessing there's a reality show out there waiting to capitalize on the idea. Or maybe a yet-to-be-made movie in Hollywood (you know, the place where we allegedly did the Moon landings).

Astronauts and cosmonauts are true professionals, highly trained and ready to deal with almost anything that is presented to them in space. But a crazed crewmate bent on murder? That's a good one and one we never simulated when I was training!

First, you would need some kind of weapon. (The Soyuz used to carry a single shotgun, but I understand that is no longer the case.) With no guns on board you would have to resort to using something else, such as a large wrench—plenty of those are in the extravehicular activity tool stowage—or a sharp pair of scissors, which can be found in the office pantry. Maybe a crew member's butter knife, which can be found in their personal eating utensils pouch, could wreak sufficient damage?

While a blunt object to the heads of your unsuspecting space fliers might be the straightforward way to "off" the crew, using a wrench, scissors, or dull knife would most certainly present a cleanup issue. Knowing how much tropical fruit punch can stain ISS surfaces, I'm guessing that blood would be even worse. Plus, being highly trained, you would know of the serious biohazard potential, necessitating the use of the ISS hazardous materials cleanup kit and following the proper procedural protocol. Ugh!

Next, disposing the bodies would be paramount. Rotting flesh, even when contained in well-ventilated aluminum modules, will still smell. Stuffing them into a soon-to-be-departing Russian Progress cargo ship should take care of that issue. After all, isn't that what they are for, sort of?

Finally, murdering all of your buddies in space would leave you with no one to dine with, no one to talk to, and no place to go. Escape paths and hiding places are extremely limited in outer space. Granted, you could extend your stay on the ISS, as your margins concerning food, water, clothing, and other consumables would soar in your favor. But I'm guessing conversations between you—the murderer—and ground control would be "touchy" at best. And you'd probably only have a couple of choices to seal your fate: either hop into a Soyuz, land solo (I don't think I had enough training for that), and take your chances with local authorities, or wait until NASA and/or Roscosmos arrange for launching something to you. I'm guessing the arriving vehicle and its posse will include new, beefy crewmates and some handcuffs!

Question: Regarding the International Space Station, how dependent is the U.S. Operations Segment on the Russian Operations Segment, given that the USOS has the solar panels?

Answer: The U.S. Operations Segment of the ISS is hugely dependent on the Russian Operations Segment. First and foremost, the USOS has no ability to re-boost the ISS orbit, which decays over time due to the constant tug of gravity, its huge mass, and the large "air" resistance that results from the very thin but always present atmosphere acting against the huge surface area of the solar array panels on the space station's outboard truss.

With the cancellation of the shuttle program in 2011, leaving all retired U.S. space shuttles in museums, the United States gave up any capability *we* had to re-boost the ISS. We now must rely solely on the thrusters of the various Russian Progress and European automated transfer cargo vehicles to provide any needed thrust capability.

You are correct that electrical power is primarily provided by the U.S. system. The U.S. segment provides much of the station's power needs and sends some of it to the Russian segment via converter devices called voltage and current stabilizers. However, the clever design of each Russian module shows that it has its own independent solar array/power system, so if the Russian segment ever decided to "undock" from the USOS, the independent station's power needs would be met by each individual module in combination. In 2007 Fyodor Yurchikin and Oleg Kotov, then my crewmates on the ISS, retracted the arrays of the functional cargo block, which is the original Russian module that was sent into orbit in 1998. I don't know if the arrays could be extended again or if they could generate power if that were the case, but I'm doubtful.

Each segment has its own ability to remove heat, to cool interiors, and to control the vehicle's attitude. While connected as they are today, they work in tandem, executing an orbital "dance" of sorts that takes advantage of the best assets of each side and maintains the ISS in a stable, power-generating attitude in the most efficient way possible. In other words, they minimize propellant and electrical power use.

In its current configuration, both the U.S. and Russian segments have their own food preparation areas and hygiene stations (toilets), located in Node 1 and Node 3 of the U.S. segment and the service module of the Russian segment, respectively. Four crew members use sleeping quarters in the U.S. segment with two sleeping on the Russian side (again in the SM). However, astronauts can sleep almost anywhere on the ISS. They need only a sleeping bag, a good place to tie it down, and sufficient air-flow and cooling (hard to sleep when it's too hot, right?). I slept in the Russian docking module for the last three weeks of my five-month stay in 2007, when new ISS commander Whitson kicked me out of my sleep station (located at that time) in the U.S. lab.

An engineering marvel in space, the ISS is a triumph of engineering efforts from more than fifteen countries around the world. Let's hope it stays up there as long as possible!

Question: Do astronauts on the ISS need to sleep with a fan on or in some way have moving air around them?

Answer: I slept in the temporary sleep station (TeSS) while living on the ISS in 2007. The TeSS, much like today's ISS sleeping quarters, had an air-conditioning inlet vent to "push" and move the air throughout the sleep station. This was critical in putting fresh air for inhalation near our faces. If this circulation were to disappear, the possibility of a carbon dioxide (CO_2) cloud in front of your face—and dire consequences—would increase dramatically. I enjoyed the cool air blowing near my head while sleeping, but I could have slept just as well upside down or sideways as the air flow velocity is high enough to "stir" the air in the sleep station adequately to ensure no CO_2 buildups.

Question: How does the night sky look from the International Space Station? To what extent, in terms of clarity compared to what we see on Earth, do astronauts have the ability to see the nearest planets, stars, and galaxies?

Answer: When asked this question in my speaking engagements, my answer is pretty simple. (I need them to be simple, so that I can understand them!) In essence, the sky looks the same to us in outer space as it does to you except you are looking through Earth's atmosphere. For the most part, we can see the same things, with a bit better clarity and more so if we take some extra steps.

Just as light pollution on the ground can destroy one's ability to gaze at the heavens, the same holds true in space. While those on Earth may head to events such as the annual Nebraska Star Party in Valentine, we need two distinct things while enclosed in a space vehicle—a window that looks to the sky, not the ground, and the ability to turn off or dim all interior lights to give us the best viewing possible.

During my five-month stay on the ISS, it was hard to get number 1. That's because nearly all of the ISS windows gazed at Earth, not at the sky. As did their docking compartment, DC-1, the two Russian sleep stations (kayutas) had windows that looked "out of plane" relative to the ISS orbit; that means kind of "sideways." So from Oleg's or Fyodor's "bedroom," or my short-term sleep compartment in Russia's docking compartment during my final three weeks in space, I could gaze at the stars somewhat. But then number 2, turning off lights, became even more important. It's hard to rid sleep stations of all ambient light so as to get the best possible view.

Today the ISS has the cupola. Imagine a glass bowl affixed to the station's bottom. While mostly focused on Earth's surface, you can look through the sides of the bowl and see beautiful panoramas of the starry sky—fantastic panoramas actually! But again, the key is to minimize or eliminate the ambient light from power supplies, computer screens, and so on. This can be done, but it takes some work—and a bunch of T-shirts.

Astronaut Don Pettit, one of the best astronaut photographers in history, began to figure out ways to reduce ambient light, using towels,

9. The Milky Way as captured in a four-second exposure from the shuttle *Discovery*'s flight deck by STS-131 commander Alan "Dex" Poindexter.

T-shirts, and anything he could to block out the light, and to make the cupola his personal photography studio. And, boy, did he ever succeed. His shots and videos—readily available on the internet—are absolutely spectacular.

On our shuttle mission to the ISS (STS-131), our commander, Alan Poindexter, was also quite a photographer. He was one of the first to capture striking long-exposure (four-second) individual shots and then, using computer software, mesh them into a video movie. One of his shots is destined to be a classic. I pushed him to submit it to *National Geographic Magazine* and have included it in this book.

Aboard the shuttle, reducing the ambient light was even tougher than on the ISS. But when you can do it successfully, the results are out of this world.

Question: What is it like to be a part of a NASA NEEMO mission?

Answer: My time living underwater as part of the NASA Extreme Environment Mission Operations (NEEMO) was unique. I was part of a real crew for the first time in my astronaut career as a member of the fifth NEEMO mission. NEEMO is what NASA calls a mission analog, or a high-fidelity simulation for actual spaceflight. By having us live underwater in the Aquarius habitat—sixty-five feet below the surface near Key Largo, Florida—NASA's goal was to put us in a mission-like scenario. We followed similar protocols to be used on the space station, including an actual mission time line.

Just like in outer space, we had specific hygiene protocols. But in Aquarius, we had something absent in outer space—a shower! Taking "Navy showers"—you turn off the water while soaping up to minimize water usage—was a real treat, ending with what Astronaut Garrett Reisman and I would affectionately call Shammy Ops (think ShamWow). We dried our bodies by first using a shammy to remove the majority of the water. Then we used our cotton towels to finish the job. Garrett and I were known to break into loud and vigorous song about our beloved shammies.

The following excerpt from my book *The Ordinary Spaceman* provides a glimpse into this highly effective analog mission concept.

10. Working on an underwater extravehicular activity construction project, clad in SCUBA gear, near the Aquarius habitat off Key Largo, Florida.

At NASA, we look for analogs through simulations and extreme environments. With the help of the great folks from the National Oceanic and Atmospheric Administration (NOAA) and the University of North Carolina [at] Wilmington, someone finally found Nemo.

Ours was not the famous clownfish Nemo of Disney movie fame but the NASA Extreme Environment Mission Operations (NEEMO), the brainchild of a group led by NASA engineer Bill Todd. An avid SCUBA diver and, at one time, a member of the Mission Operations Directorate training team, Bill and his cohorts came up with a fantastic opportunity for station crewmembers to "get their feet wet." We would learn about living and working on the space station in an analog below the ocean's surface, a barnacle-and-coral-encrusted habitat anchored near the deep coral reefs of Key Largo, Florida.

We were known as NEEMO 5, the fifth crew in this brand-new experience. Everyone—the crews, the mission support team, the NOAA divers leading our training, and the scientists—was essentially learning on the fly. The idea was simple and elegant: put potential station crew members together underwater in an environment hostile to humans, then have them execute a mission as similar as possible to an actual one on board the space station.

We would have mission tasks to execute, a timeline, extravehicular excursions outside the safe confines of "home" to perform maintenance and repairs, all while gathering scientific data and testing new concepts and equipment applicable to the undersea environs. We would be working in tandem with a mission control team, in near constant communication during our time under the sea. Some of us would even be forced to deal with the concept of family separation, a key component in the life of a crew member experiencing long-duration spaceflight.

Our experiences in South Florida commenced with a week of dive training to help us understand our equipment as well as the danger posed by our home beneath the sea. Under the tutelage of seasoned water lover Mark Hulsbeck, appropriately nicknamed "Otter," we worked on enhancing our SCUBA skills, got briefings on the science we were to use and execute, and made sure that our personal items were in order for the trip to the ocean floor.

Life was good in the Aquarius habitat, positioned "20,000 millimeters under the sea" and located about four-and-a-half kilometers off the shores of Key Largo, in Florida's National Marine Sanctuary.

In this near-perfect space flight analogy, I loved to relax on my bunk bed and listen to Enya's beautiful "Caribbean Blue." From my perch on the habitat's port (left) side, I could gaze out the aft window and watch spotted eagle rays, Caribbean manta sting rays, and nurse sharks silently and smoothly cruise the ocean floor. . . .

The Aquarius habitat was our home for two weeks, about as long as a shuttle flight (usually eleven to fourteen days). Six of us—three astronauts, one scientist, and two habitat technicians ("hab techs")—comprised the crew for our NEEMO mission. The size of our crew, as well as many of our activities inside Aquarius, paralleled those planned for the space station. We dined on actual ISS cuisine. We tested equipment destined for the station, some of which I would eventually get to use in space.

The work kept us extremely busy, especially the first few days. I felt constantly rushed, like I didn't have enough time to do everything I needed to do. As it turned out, I felt the same on my long-duration space flight. Typically it takes the crew a couple of days to establish their individual, and then collective, routines.

Mostly rookies, our crew was led by veteran astronaut and native Iowan Peggy Whitson. Peggy was formerly the chief science officer and flight engineer on the fifth excursion to the ISS, and was well suited to teach and mentor us so we could one day quickly adapt to the zero-gravity environment of an eventual space station assignment.

Question: What is it like to eat astronaut food on a regular basis in space?

Answer: Surprisingly, I found astronaut space food to be quite tasty, especially the Russian versions. Russian space food contains more salt and fat content, therefore making it tastier than the American versions in my opinion. During my five-month stint on the ISS, I ate more Russian than American food, while Fyodor and Oleg preferred American dishes for some reason. That suited me just fine. Oleg and I did try some Indian food left behind by Suni Williams. While I'm unable to recall the exact dishes (one was macadamia nuts in some kind of red sauce), I do recall that neither of us was too impressed! Finally, when I flew on STS-131, I found that shuttle food just gave me gas. My ISS combo of Russian and American food, though, kept me as regular as clockwork!

Question: Is it possible to isolate a body part's exposure and reaction to a zero-gravity or reduced-gravity environment? For example, do astronauts have to do deep vein thrombosis prophylaxis/blood clot prevention before going on or during space missions?

Answer: Nope, there is no way to do that. Your entire "innards and out-tards" are all along for the greatest ride in the universe! And while I guess we could do exercises similar to those that folks do on long-duration airplane rides, I didn't do anything so detailed as what you reference. But I was told prior to my first liftoff with STS-117 on *Atlantis* that I could get a massage before launch. Turns out it entailed just a couple of our astronaut trainers giving me electrostimulation therapy on my lower

11. Enjoying some Russian chow in the Russian service module Zvezda.

back. Although my fantasy of a gorgeous Swedish masseuse did not materialize, the electrostimulation did relax my lower back such that I had minimal problems when I reached orbit a few hours later.

Question: What is it like to float in space?

Answer: This is one of my all-time favorite questions, leading to my all-time favorite answer: I was Superman every day! I flew to breakfast. I flew to work. I flew to the bathroom, and I even flew while I was *going* to the bathroom! And like Superman, I was faster than a speeding bullet (we flew around the Earth at 17,500 miles per hour, or 5 miles per second). I was more powerful than a locomotive (I could lift anything on board the ISS with one hand). I was able to leap tall buildings in a single bound (no buildings are taller than I was at 225 nautical miles above Earth). And I fought for—as I still do today—truth, justice, and the American way!

Question: Did shuttle astronauts sleep in the space shuttle when it was docked to the ISS?

12. Living on the ISS, I was Superman every single day.

Answer: The answer is yes and no. There are limits as to how many crew members may sleep on board ISS, as its systems are designed to accommodate a certain number of living, breathing, sweating humans. Preflight, the ground control team gave permissions as to how many could stay on the ISS for sleep periods. Then the vehicle commanders determined which shuttle crew members could sleep and where on the ISS. Usually the shuttle commander stayed on the shuttle, remaining ready for any possible alarms or situations that could crop up during the night. During STS-131, Rick Mastracchio and I slept in the ISS airlock as we would be spending many hours working there, prepping for our three space walks. When I lived on the ISS for five months, we often had various combinations of shuttle crew members camping out in the Node 1 and airlock, as those were the two most comfortable places to "crash."

Question: Can we see the ISS from Earth?

Answer: Why certainly we can see the ISS from Earth. We just have to know when and where to look.

As noted in chapter 2, the ISS orbits our Earth every ninety minutes due to its current positional orbit about 250 nautical miles up (we flew at about 225 nautical miles in 2007). Since it completes multiple orbits in a twenty-four-hour period (okay, students, do the math!), you have ample opportunities to see the ISS pass overhead as long as your location is beneath its ground track. The ISS is pretty easy to spot because it is so large, about the size of a football field. When the sun's rays hit the station, for all intents and purposes, it lights up, becoming the second-brightest object in the night sky behind our Moon. Not bad company to keep in the annals of brightness.

But the Earth also rotates on its axis every minute of every hour of every day. That means—in the most simplistic sense—the space station's orbit shifts relative to that rotation. Don't worry, though. NASA has figured all of it out for you and created smartphone applications and websites to assist you in making successful attempts to "spot the station." Search your device for the apps or go to NASA's "International

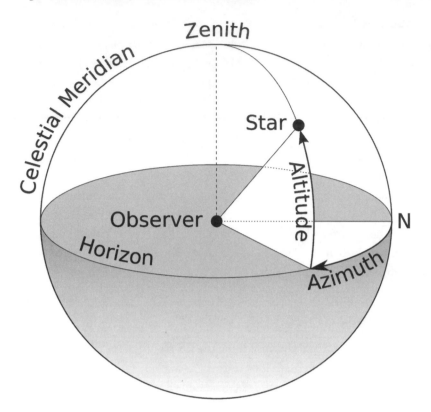

13. Spotting an orbiting spacecraft requires you to know when and where to look. If you can find north, and have the object's azimuth and elevation, you're ready!

Space Station" page (http://www.nasa.gov/mission_pages/station/main /index.html) to find when the ISS is next flying over you.

Now you have to know where to look too. This is done using what star gazers refer to as "azimuth" and "elevation." Whoa, rocket science. Don't sweat it though, as the NASA website and the apps have a nice explanation. But for the lazy ones out there, like me, the website and apps will tell you the time, the direction (the azimuth) from which ISS will rise and then set, and its elevation, or how high it will be (and how long) above the horizon.

Good luck on your ISS hunting expedition. This is the *perfect* time for me to say, keep lookin' up!

Question: What are the languages spoken in the ISS?

Answer: Many languages are spoken on the ISS. At times one could probably have heard Japanese, Italian, French, German, Russian, English, and even Portuguese. However, by decree resulting from mutual agreements, the official language of the ISS is supposed to be English. In other words, all crew members are to be fluent in the English language. With that being said, that's not exactly how it works.

Since the Russian Soyuz rocket/capsule is now the only way to lift astronauts and cosmonauts into space, the Russians have a tremendous amount of control. In this position the Soyuz serves as the only ISS lifeboat, so it has always been required we speak and read Russian in order to ride along. All Soyuz procedures, labels, and so on are written in Russian, dictating U.S. and international astronauts must reach a level of fluency in that language. During my training before 2007, we quantified it as level 3. This is on a scale of 1 to 5, where 5 describes someone who is fluent in the language.

Contrast that with the situation in the U.S. and Russian segments of the ISS, where all procedures and labels throughout are written in both English and Russian based on a compromise reached many years ago. This was established long before crews had ever reached the confines of the orbiting laboratory, but it is not in total compliance with the ISS international language agreement.

These requirements for American astronauts to speak, read, and understand Russian at very high levels add many weeks to an already long training flow. Unlike many other countries, where English is taught to children as a second language at an early age, most Americans don't speak Russian as a foreign language at all. That learning the Cyrillic alphabet is involved further complicates the matter.

But most astronauts are up to the challenge and study hard to learn Russian to the best of their ability. After all, their lives may depend on it.

Question: Does the structure of the ISS ever creak or groan? If so, is it because of gravitational differences?

Answer: Yes, the ISS makes noise all the time. Those noises *may* include creaks, groans, hums, buzzes, beeps, clicks, and squawks from the various pumps, fans, switching valves, and computers. Loud bangs, such as when spacewalking astronauts are crawling around on the outside, can be heard on the inside, but they indicate our crewmates are working and alive. Sometimes, it may not smell too good either, but we mustn't go painting an inaccurate picture for those who dream to live and work there one day soon.

My NASA colleague and good friend Robert Frost likes to point out that thermal cycling—that is, the rise and fall of the temperature in space due to the sun being blocked by Earth during our ninety-minute orbits—causes the space station's structure to flex and its metals to expand and contract. It is analogous to a car or motorcycle with an aluminum engine that makes a ticking sound as it cools.

Ambient noise inside the station is comparable to that experienced from the lower decks of an aircraft carrier, or somewhere in the fifty-decibel range if my memory serves me correctly. Essentially it is "white noise," which can be offset if needed by foam earplugs.

While sleeping, I often felt as though I was hearing the constant rotation of the station's solar arrays as their solar alpha rotary joints slowly turned to keep the arrays facing the sun. More than likely I was sensing the vibrations through the station's hull. However, as someone who lived there for 152 days, I am here to tell you that for me, it was an *excellent* place to get good rest. (I averaged seven hours and twenty minutes of sleep a night during my stay.)

Question: What is the total square footage of space on the ISS? How large is the largest living space by volume on the ISS? How many people can be accommodated on the ISS at one time?

Answer: I don't know the total square footage of space on the ISS. I *do* know that we should be talking three-dimensional volume vice square footage. In the microgravity environment of low-earth orbit, we are able to utilize each and every direction—forward/backward, right/left, up/down—of our sequestering habitat, unlike what we can do here on

14. The ISS is huge—about the size of a soccer pitch or an American football field.

Earth. (You do not have up/down capability unless you're a bird, a plane, or, um, Spiderman.)

The ISS is huge—as noted, about the size of an American football field or, for you soccer lovers, a soccer pitch—and weighs nearly a million pounds. It stretches about 120 yards from starboard solar array to port solar array, and the arrays are about 60 yards tall. The interior pressurized volume of the ISS is estimated roughly equivalent to that of a Boeing 747 jetliner, or about 32,333 cubic feet. Its size is similar to that of a six-bedroom home, except you can use the space near the ceilings.

Today's ISS has two bathrooms (in 2007 we only had one in the Russian SM) and two areas for food preparation—one in the Russian SM and another in the U.S. Node 1. It can support a nominal crew of six, with an ability to handle around a dozen folks in the days of shuttle cargo deliveries for periods of a couple weeks.

Its solar arrays produce ample electricity to operate its systems. Roughly it is equivalent to the amount of juice needed to power about fifty three-bedroom homes.

Today, it flies about 250 nautical miles above Earth, or about the distance from my home town of Ashland, Nebraska, to North Platte, Nebraska. At its orbital inclination—or the angle it flies relative to the equator—of 51.6 degrees, its path covers about 80 percent of the Earth's surface as it orbits our planet sixteen times a day. This is a very good thing operationally, as it flies periodically over all the nations involved in

Pressurized volume: 1,204 cubic meters
Accommodate 7 person crew

Power: 80 Kilowatt (KW) average
with 40 KW average for research
(Apprx. equivalent of 50 homes)

15. The volume inside the ISS is about the same as that of a Boeing 747. Its solar arrays generate enough power to sustain fifty U.S. three-bedroom homes.

its construction and operation—specifically, the United States, Russia, Canada, Japan, and the member states of the European Space Agency. Composed of multiple modules of Russian, U.S., Japanese, and European origins, we can guesstimate that most are on the order of 60 feet long and have a 15-foot diameter. That's because many of them were delivered to space by the space shuttle, whose payload bay easily handled those cargo dimensions.

Then multiple racks are placed into these modules, a curved side against the station's outer hull, with a vertically flat front exposed in the interior of the ISS. This leaves us with a roughly square corridor of living space—remember, we operate in three dimensions—in the center of each module.

The number of inhabitants at any given time on the ISS varies. From the initial Expedition 1 crew of three people, the size has ranged from two (after the *Columbia* accident, crew size was reduced for a while as a precautionary measure) to a recent group of nine to thirteen spacefarers, who were all happy as clams aboard our engineering marvel.

During my time on board in 2007, our three-man crew—Kotov, Yurchikin, and me—grew all the way to ten people when visiting shuttle crews arrived. As STS-131 ISS visitors in 2010, our seven-person crew pushed the total to thirteen inhabitants when we added to the then pretty standard number of six crewmates on board at one time. Personally, I preferred three. Thirteen people aboard were way too many folks.

5

Space Physiology and Psychology

Question: Do astronauts get spacesick? While it wouldn't be the same as being carsick or seasick—as they're due to motion sickness—zero gravity must do something to the human body, which has evolved over the years in Earth's gravity.

Answer: Many astronauts get spacesick, but it's usually a temporary thing. Using your words, to me it's *very* similar to being carsick or seasick. It is related to motion sickness, as the body's vestibular system adapts to weightlessness. In nearly all cases, if you have it, then you're likely to throw up.

I don't know the exact, quotable percentage anymore, but about one-third to half of all astronauts flying in space experience some form of space sickness. NASA flight surgeons refer to it as space adaptation syndrome—of course, we have to have some fancy term for it—and it ain't fun; but I wouldn't know through experience. Honestly, I *never* got sick while in space. But I did take appropriate medications on launch day to help me avoid the syndrome. After all, your tax dollars were invested in NASA, and America was counting on me to perform at the top of my game while in space. If meds would help me, and apparently they did, then I was going to take them.

My first launch on STS-117 was with a crew of veteran astronauts and four other rookies. A high percentage of our crew could be found with white bags tightly positioned around their mouths at various times during the first two days in orbit. Fortunately for most, our bodies adapt quickly, and the third day often brings relief and a normal appetite to even the most spacesick-haggard astronauts.

Astronauts have been known to "blow chunks" just from flipping upside down. In not maintaining a normal, earthlike heads-up attitude, our brain can flip out a bit, causing many of our steely-eyed heroes to give a "technicolor yawn" for their crewmates. My need for the white bag did not occur until landing day and after a serious session in the port-o-potty.

Question: How does our body adapt to weightlessness? What are some concerns in the microgravity environment?

Answer: There are definitely concerns for humans living in a microgravity environment. For example, our immune system performance seems to degrade a bit while we live in outer space. Fortunately, our human body—led by our brain—is an incredible thing. In just days, it begins to figure out the nuances of our strange new environment, adjusting our human systems to function at near-peak performance. And we are executing numerous experiments on board to determine exactly why these things occur and how we might compensate for that in the future. The information we glean will hopefully lead us to more effective countermeasures for those problems on Earth as well.

Another primary concern is musculoskeletal degeneration. That's why astronauts exercise for up to two and a half hours every single day, using aerobic and resistance training methods, and take supplements such as vitamin D tablets.

If astronauts aren't careful (and we are very, very careful), the International Space Station could be a great environment for bacterial and fungal pathogens. But we periodically test the environment for them and, if any issues are found, take corrective action. To date, there have been few, if any, concerns stemming from the station's internal environment and the potential for bacterial companions to our brave spacefarers.

It has been said you should think about all this when you say you want to go to space, as it makes life aboard a submarine look like a picnic. Well, I have visited both a submarine—the Ohio class ballistic missile submarine dubbed the USS *Nebraska* (SSBN-739)—and the ISS, and I highly prefer life on board the ISS. It has way more space per person, and I didn't have to sleep behind a Trident nuclear missile.

Question: In astronaut training, astronauts ride on a huge centrifuge to make them experience high accelerations. Is this only to test how their bodies react to such large gravitational forces?

Answer: During shuttle training, we went to a U.S. Air Force base in San Antonio for a single ride in its centrifuge. We "flew" the shuttle's ascent and entry profiles to give us the exposure to what two to three g's would feel like. Actually it was no big deal, and we were just "checking a box" on our preparation list.

At the Gagarin Cosmonaut Training Center in Star City, Russia, I flew in its centrifuge as well. Since I was a rotating expedition crew member via the shuttle, I only did the Soyuz reentry centrifuge profile in a manner similar to that of the shuttle, pulling the requisite number of g's at the appropriate times. However, we also did some separate runs that reflected a ballistic Soyuz reentry profile. This reentry, a contingency, is extremely dynamic. We pulled eight and ten g's for short periods, reflecting what we would experience in the event of a failure driving us into that mode. It is extremely interesting to note that my crewmates Fyodor Yurchikin and Oleg Kotov of Soyuz TMA-10 and then Peggy Whitson and Yuri Malenchenko of Soyuz TMA-11, in sequential returns to Earth, experienced Soyuz ballistic reentry trajectories due to separation failures between the spacecraft's orbital module/living compartment and the reentry module/capsule.

The third member of Malenchenko's crew—a South Korean space flight participant—sustained back injuries as a result of this reentry. She battled them for quite some time following the mission.

Question: Does the ISS have a blood bank for astronauts in case of any emergency?

Answer: Great question, and the answer is no.

If an emergency is serious enough to require blood, astronauts on board the ISS have little recourse. The station does not have the capability to store blood in quantities to be used for transfusions. Blood samples taken from the astronauts are stored in a freezer—called the

Minus Eighty-Degree Laboratory Freezer for the ISS—that can keep things at minus 80 degrees Celsius.

The two Soyuz capsules always attached to the ISS serve as lifeboats for the entire crew. Each crew member must take the designated Soyuz that brought him or her to space. It is where the astronaut's Russian Sokol (ascent/entry) suit will be stored, and that's the one he or she will hop into if an emergency requires a quick return to Earth.

For serious emergencies, the plan is for the crew to stabilize the injured crew member and then do an emergency undocking and return to Earth. The crew can be on the ground in roughly three and a half hours from the time the Soyuz leaves the ISS.

Here's hoping we never have to invoke that capability.

Question: Does the ISS have an on-call doctor?

Answer: Well, my answer is yes and no. It all depends on your definition of "on call."

The ISS Mission Control Center in Houston has a staff of flight controllers available 24/7, 365 days a year. Each day one of those flight controllers also carries the moniker of "flight surgeon." That person serves as the medical eyes and ears for every astronaut—including cosmonauts, if needed—on board the ISS. But the flight surgeon is now 250 miles away and seated at a computer terminal. His or her ability to aid the crew is limited in many aspects.

So for each mission, NASA chooses a crew medical officer, which was one of my roles when I was a member of the ISS Expedition 15 crew and also during my STS-131 mission on *Discovery*. In that capacity I was able to provide *limited* support to any onboard medical situations or emergencies. During my ISS mission, fortunately for my Russian crewmate Yurchikin and me, our third "space brother," Kotov, was a trained medical doctor, as in ". . . turn your head and cough."

As a crew's medical officer, I received special—and additional—training for handling several key scenarios. For example, I could be counted on to be able to collect blood samples, perform urinary (Foley) catheterizations (this has actually happened in space but not on my missions), execute defibrillation (perhaps the use of the word "execute" is ill-advised here?), perform intubations, administer sutures, and provide basic dental (pull a tooth) and eye care (search for foreign objects). I even did a stint in the emergency room of a Houston-area hospital where I actually sewed up a patient's leg that was injured in a four-wheeler accident, administered a shot during a drug addict's recovery, intubated a patient, and performed two Foley catheterizations. I also watched and then pointed out (that's my style) how two interns had botched the administering of a patient's anesthetic. They were not very happy with me, a lowly astronaut. It was a tremendously interesting night for me, and, fortunately, no one died.

So just how do we handle onboard medical situations? If serious events arise, we contact the ground control team immediately, assuming communications are normal, while heading to our medical procedures book. Through our ability to talk directly to the flight surgeon in real time, we are able to confirm and then work through the appropriate procedures. (Note here that being able to access help in real time will be a *huge* issue on a mission to Mars, with an anticipated twenty-minute communication delay both ways.)

To date, thankfully, we have had no critical medical issues on the ISS, but when we do, it will be interesting—and perhaps nail-biting if it's serious—to see what the outcome is. For example, consider a heart attack or burst appendix scenario. It is very possible that part of the crew will quickly depart the station for Earth. And if that were the case, how hard would it be to stabilize the patient, put him or her in the Sokol space suit, strap him or her into the Soyuz, depart the ISS for Earth, land, and get to a hospital? These are very scary considerations for me.

What's the bottom line then? We don't want these scenarios to ever occur. This is why astronaut candidates are required to go through so much medical screening before selection as astronauts and then again before they are assigned to fly in space. But one day, when humans actually move away from our secluded planet Earth, we will need to be ready to accept the challenge.

Question: Are there any recorded incidents of astronauts/cosmonauts having a panic attack in space?

Answer: When I accidentally squirted butterscotch pudding all over the right arm of my Russian commander in 2007 during dinner, I nearly had a panic attack. Turns out it was all for naught as Yurchikin loved butterscotch pudding.

One story does exist from the space shuttle days that describes a rookie crew member who struggled to adapt within the cozy confines of the shuttle's crew cabin. According to the tale, that person was apparently battling claustrophobia to some degree and attempted to open the shuttle hatch. Now *that* could have had disastrous results. During my two shuttle flights, I was given the responsibility to put the hatch-locking mechanism in place once we reached orbit, so perhaps the story was true after all.

Question: When closing his or her eyes, why does an astronaut see "lightning"?

Answer: Astronauts don't see lightning when they close their eyes. The high-energy particles speeding throughout our solar system sometimes make it inside the space station and into an astronaut's eyes. I did experience those "flashes," as do all astronauts. For me, the frequency of their occurrence seemed pretty low, but I knew when I had one, especially just as I was trying to fall asleep. As I headed off to dreamland, if one of the particles excited my optic nerve, the flash was pretty noticeable. Hopefully, no permanent damage was done.

Question: I'm color-blind. Does this ruin my chances of becoming an astronaut? If one can see all the colors of the rainbow, but is challenged to distinguish certain tone patterns when arranged in a certain way, is that allowed?

Answer: Unfortunately, being color-blind—at least in my understanding of the current astronaut selection criteria—does eliminate you from consideration.

As an astronaut, many of the things we need to know and understand are related to color. From our caution and warning system, which uses *red* for "emergencies/warnings" and *yellow* for "cautions," to our simple packing system—which uses *green* to indicate "return to Earth," *yellow* for "send to outer space," and *white* for "permanent location"—colors are a part of our daily lives in space. Not having the physical capability to discern these colors quickly and clearly could impair a crew's ability to respond appropriately and rapidly.

Rumor has it one astronaut was able to "beat the tests"—perhaps it was before being color-blind was a disqualifier—and was admitted into the astronaut corps even though he was actually color-blind. If true, that case was certainly an exception.

But don't despair. The way technology is advancing these days, you never know what may happen in the world of ophthalmology. It's possible that one day your inability to see in color could be fixed.

With regard to color tones and pattern recognition, I don't have a great answer for that. My guess would be that those issues would severely limit your chances. However, you could submit an application and take some preliminary eye exams to see if you meet the criteria.

Question: Is it true that staying in space for months makes your eyesight worse?

Answer: Apparently, the answer is yes, for some anyway. Please note that I, the Ordinary Spaceman, did *not* experience this eyesight change. It appears from additional preliminary NASA studies performed at the JSC in Houston that I am fortunate to have a special protein sailing through my body that did not allow this phenomenon to occur. Further analysis is ongoing to nail down the causal specifics of this deterioration and how to prevent it from occurring. The overwhelming thought at this time is increased pressure on the eye's optic nerve—given the movement of body fluids into the head resulting from microgravity—is causing a significant change in some astronaut's near and distance vision.

Keep lookin' up—that is, as long as your eyesight is still good!

Question: Can astronauts see Earth's lights from space?

Answer: Yes and no! Astronauts can definitely see Earth's lights from space. What you need to know is that for any given orbit, roughly two-

16. Today's camera equipment allows astronauts to capture amazing pictures of
Earth at night. Here we see the city of Dubai in the United Arab Emirates.

thirds of the Earth's surface is in sunlight, while the other third is in
darkness.

Think of a flashlight shining on a basketball. While the light may shine
directly on the forward portion of the ball (or our globe), some light is
able to reach around to the right and left, top and bottom, leading to
the rough estimate of two-thirds.

The pictures being taken by ISS astronauts these days are spectacular, due in great part to the wonderfully modern cameras on board. A
far cry from the Kodak DCS 760XD I used in 2007, the Nikon D3S (hope
that moniker's correct, Nikon), coupled with some new and powerful
lenses, are yielding amazing photos.

Question: What's it like to exercise in space on a regular basis? Is the
fitness equipment on the ISS used by all astronauts, including those of
the Russian Space Agency?

Answer: I enjoyed my exercise time on the ISS immensely. While serving to keep my muscles and bones in condition, it was also a great stress
reliever for me every day. I exercised most mornings after breakfast for

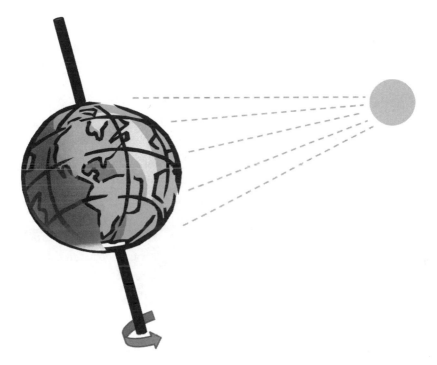

17. As the sun shines on Earth, part of the planet is in total darkness. Courtesy of the author.

two and a half hours a day, using the treadmill; the interim resistive exercise device (IRED), which is a cable/pulley–type weight-lifting device; and the stationary bicycle, which is known as the cycle ergometer with vibration isolation and stabilization. While running on the treadmill, I watched movies via laptop using DVDs sent up to the station by the folks on the ground. (I got through all of the Harry Potter flicks that were available at the time. I was hooked!) While using the IRED, I watched the NBC *Nightly News* with Brian Williams and my favorite sitcoms: *Hogan's Heroes*, *Get Smart*, and *The Beverly Hillbillies*. Riding the bike, I watched current TV programs such as *Law and Order SVU*. All of these files were uplinked by our crew support personnel. They would record the shows, then convert them into computers files that could be sent up overnight. Quite a tedious process, but I was certainly grateful!

Exercising in outer space was similar to exercising on Earth, with the major difference being no gravity. Given that major delta, I required harnesses to hold me down on the treadmill and during weight-lifting exercises such as squats and lunges. Special biking shoes with pedal clips allowed me to cycle to my heart's content.

Perhaps the best part of exercising was the psychological boost I got from the cleanup. While no shower was available, a good towel bath and some fresh clean clothes went a long way in picking up my spirits and getting me ready for the workday ahead.

From personal experience, the crew of Expedition 15—Yurchikin, Kotov, and I—all used the exercise equipment during our five-month stay in outer space. I worked out nearly every single day. The Russians had their own bicycle-type ergometer, and the treadmill resided in their segment. Their ability to use the U.S. equipment was negotiated through the various management teams involved with all aspects of the ISS international relationships, so I do not have intimate knowledge of what those arrangements were or how they came about.

Today crews enjoy the benefits of the new advanced resistive exercise device, an updated version of the IRED that uses hydraulic pistons to create the resistance rather than the IRED cable system. Some astronauts are coming home *as strong or stronger* than when they left Earth. The **C**ombined **O**perational **L**oad-**B**earing **E**xternal **R**esistance **T**readmill, named after the politically spoofing pundit Stephen Colbert, provides their aerobic training as does the exact same stationary bicycle I used back in 2007.

Stay fit, America!

Question: Do astronauts ever skip a day of exercise while on the ISS? What's the longest known time that an astronaut has not exercised while in orbit?

Answer: I can only give you my answer to the question of skipping a day of exercise. I do not know about other astronauts, and even if I did, I probably wouldn't rat them out! I'm guessing some astronauts have gone on the order of days without exercising, but it's certainly not in their best interests.

18. Performing some single-leg squats using the interim resistive exercise device, 2007.

Of the 152 days I spent in space aboard the International Space Station in 2007, my first two days were on the space shuttle *Atlantis*, which was my ride into orbit. My final two days in space for that mission occurred with the crew of *Discovery*, which was my ride back to Earth. I performed my requisite exercise protocol religiously all four days on the shuttles by using their stationary bicycles.

During the 148 days in between, while living on the ISS, I skipped my prescribed exercise period only once. Although expected to exercise two and a half hours every single day on the station, it was a Sunday, and I decided I needed to take a break. Boy, did I regret it.

It turns out exercise was more than just physical exertion for me. Sure, it helped to keep my bones and muscles strong in preparation for my return to Earth's gravity, but the psychological aspect of exercise was perhaps much more important.

When I took that day off, I didn't feel right. I felt as if I had cheated myself of something very important. While my intention was to give myself a well-earned and much-needed respite, it turned out to have the opposite effect. I *needed* exercise and what it provided.

As I have said, perhaps my favorite time of day was the cleanup after exercising. Seriously, I loved how I felt after getting my body clean and then donning my clothes. Having finished with exercise for the day, I could then concentrate on the tasks ahead. Imagine my elation when the day included donning new, fresh out-of-the-bag clothing. What a rush!

Question: What is it like to be a physical trainer for astronauts?

Answer: The Astronaut Office does have professional physical trainers. In true NASA acronym fashion, they are called ace-ers, for astronaut strength, conditioning, and rehabilitation (ASCR). But, believe it or not, a majority of the astronauts do not use their services or even the astronaut gym itself. Why? It's hard to say, but many of them don't believe they are needed or that they know more about their training needs than these professionals do. As a spacewalker and lifelong athlete, my opinion is these experts are desperately needed and valuable.

Many astronauts have injured their shoulders while training for space walks due to the poor design of the current extravehicular mobility unit. Once injured, many have required shoulder surgery, after which they *have* sought out the trainer's expertise to execute their rehabilitation. One might ask the question, If they would have used that expertise prior to the injury, could the injury have been prevented?

In today's long-duration space environment, the trainers are more valuable than ever. Aside from preflight physical preparation, the typical rehabilitation of an astronaut after a six-month mission can take as long as six weeks to bring the astronaut back to "normal." And that does not include the *complete* recovery of lost bone and muscle mass. Much of this recovery time is based on how hard the astronaut is willing to work and how hard he or she worked prior to leaving the planet.

The NASA astronaut trainers are wonderfully knowledgeable, highly professional, and a whole lotta fun. Many of them served at major universities—Iowa State, Michigan State, and the University of Nebraska, to name a few—while working with highly tuned athletes. They are all NASA contractor employees at this time, and recent budget cuts continue to threaten their numbers at the Johnson Space Center. Obviously, I am a huge proponent of their work.

Question: What logistical challenges must astronauts overcome when having a bowel movement in space?

Answer: This is what I call dealing with the "void" of outer space! Again I turn to *The Ordinary Spaceman*:

In performing "number two"—pooping, defecation, crapping, sending the Browns to the Super Bowl—the . . . astronauts learn early on in their spacefaring careers whether they are "simo" (for simultaneous) or "serial" performers in the toilet. Serial is best (in my humble opinion) as it gives you the opportunity to concentrate on one thing at a time, eliminating the need for multitasking.

Assume that we are serial performers and our personal pre-mission process analysis has shown we urinate first. Following the successful conclusion of our efforts to "tee tee," . . . we may proceed to the more "concrete" activity of pooping.

It is *extremely* critical for everyone to have accurate aim. Imagine yourself on a camping trip, ready to perform this critical bodily function, but your target is a hole about the size of the opening on a jar of spaghetti sauce. Further imagine that when you do poop, it won't fall to the ground. It won't fall at all. Gravity is essentially absent and there is no "separation factor."

Without gravitational pull the process of executing number two relies on near-perfect body position and flawless technique.

My preference (learned only after a few bouts of trial and error) was to totally remove my lower body clothing and stuff it behind the handrail opposite the "throne." That made it easier to position my cheeks to maximize the possibility of success.

The ACY [Russian] toilet is very much like those used for camping: an empty can with a hole in the top. A small plastic bag with tiny holes in its bottom is inserted into the hole and attached to the rim of the can with an enclosed rubber band. . . . [System fans provide a small tug that fully extends the plastic bag (by virtue of those small holes) and sucks] the nasty odors into the system where they are effectively absorbed by filters.

If our aim is successful and our diet is sound, . . . [what leaves us makes] its way into the very bottom of the bag.

The fan's pull in the toilet is . . . [small], making it highly unlikely the . . . [poop] can break free on its own. Separation factor is key. By donning a rubber glove and executing a scissoring motion with your index and middle finger, you slice through the offending feces and gently push it down where it's supposed to go. . . .

I did investigate alternative techniques to the "scissor," . . . [including] the "wiggle" . . . in a move akin to a Miley Cyrus twerk. . . .

Cleanup operations are as on Earth. Personally, I did not use normal toilet paper, as Huggies wet wipes and Russian gauze pads sufficed nicely.

The next step was disposal of the rubber glove, pulling carefully from the wrist to turn it inside-out, keeping everything of importance squeaky clean.

The final act was pulling the [bag's] rubber band free of the waste can's rim and giving the now-sealed bag one last push into the can. . . . [A] second Huggie, followed by a disinfectant wipe, was the main way to defeat uncleanliness in the "post-op" phase.

Question: How often do astronauts on the International Space Station go to the toilet? Because of both changes in digestion and foods eaten, do astronauts not have to go to the bathroom as often as we do on Earth?

Answer: Not to be a turd or anything (get it?), I disagree with the suggestion we don't have to go to the bathroom as often in space as we do on Earth. Not that I (or anyone else for that matter) specifically counted the number of times I executed number 1 or number 2, but my primarily Russian food diet kept me quite "regular" in the bathroom department, much like here on Earth. (Oleg and Fyodor loved American food, and I preferred their Russian cuisine.)

While living in space for 152 days as a member of the ISS Expedition 15/16 crew, I found that the composition of my food intake—combined with weightlessness, regular exercise, and so forth—allowed bowel movements to occur when everything was "ready." I am not trying to be funny here; it's just hard to describe.

You see, on board the ISS, I never had any question about when I needed to perform number 2. It seemed that each trip into the Russian toilet system was a complete purge of my system. I guess this happened one to two times per day.

Russian specialists determine the quantity of supplies needed for "bathroom ops" based on strict numbers of bathroom trips per day per astronaut or cosmonaut divided by the mission length (assuming various intakes of food and drink). Just as here on Earth, my experience with urination was that the more liquid I drank, the more I had to pee. Not exactly rocket science.

I do not recall ever discussing in any briefings—U.S. or otherwise—that we would or would not experience differences in our bodily functions. Perhaps I was in the bathroom during that part of the lecture?

Question: How does the human waste removal system on the ISS work?

Answer: A simple yet technical answer can be stated as follows. There are two interfaces used for the toilet. Urination is performed by directing the fluid into a hose with a yellow funnel on the end. As noted previously, defecation requires the astronaut to position himself or herself directly over the commode and its circular opening. Once lined up properly, activated airflow will direct the feces to go in the right direction and into a collection tank upon leaving the body.

While liquid waste is recycled into drinking water, solid waste is stored in the replaceable collection tank. When full, the tank is put into a Russian Progress cargo vehicle, which will burn up in the Earth's atmosphere during reentry.

The not-so-technical (and way more fun) answer? Solid waste removal on board the ISS is a bunch of crap. That's right. You heard me: a bunch of crap. That's because it accumulates.

Pooping in space (and microgravity) can be considered an art form. As noted previously, it's similar to going while on a camping trip, except we make sure that you poop into a can—specifically, a Russian-built aluminum can. It's two pieces of aluminum, and they need to be assembled once in orbit. With the appropriate accoutrements and connections attached to its top, and the can snugly affixed to the floor, it takes its place prominently in the space bathroom much like the throne in your home on Earth. (Today's ISS has two bathrooms, one in the Russian end and one in the U.S. end; please don't read anything into my use of the word "end.")

Prior to entering the waste into this large can, the poop first goes into a small plastic bag with holes in its bottom and with an enclosed rubber band around its top. The holes let air be drawn through the bag, pulling any smells down into the system filters. (This is a good thing.) In addition, the air flow through the holes inflates the bag, making it much easier for your poop to reach its intended destination. We don't want to imagine what happens when you miss. (I did this on orbit but only once. It was not a good day.)

After completing a bowel movement in space, where does the poop go? After cleaning up, all the waste is stuffed into the plastic bag with the already-present fecal matter. The bag is then liberated from the rim of the can (this is what the rubber band clings to)—it snaps shut—and the plastic bag goes down into the can to float freely in the absence of gravity until many more plastic bags join it.

Each can is designed to hold a specific amount of human waste, measured by individual bowel movements. A person is essentially calculated to have a couple of bowel movements per day. When the can approaches being full, depending on which bathroom location either a Russian cosmonaut or an astronaut dons a rubber glove and pushes the poop down, compacting it as much as possible to allow a few more bags to fit. When it's finally full, the can is sealed at the lid—two wing nuts do the job—and is temporarily stowed in an area in the Russian service module. When packing begins, the full aluminum cans are one of the first things to go in the Progress cargo vehicle.

Keep lookin' up, and hope you don't see a piece of space poop barreling toward you at orbital velocity. If you do, duck!

Question: How do farts behave in low gravity? I know this sounds almost embarrassingly infantile, but I had to ask when the following thought crossed my mind. In low gravity, as in the ISS, there is barely any convection of gases, so the gases that are released should take far longer to dissipate and be filtered out. Am I right in this theory? Does this lead to awkward situations?

Answer: Wait. Ahhhh, I feel much better!

As a key contributor to the zero-gravity world of flatulence—having lived on board the ISS for five long months in 2007 and with a twelve-day return mission in 2010—I now consider myself somewhat an expert.

Flatulence is a normal process, even in space. It is an indication of "things to come," just as here on Earth. When your gas starts to stink and is inhaled by those around you, manners dictate that it's time to find the toilet and perform a "waste dump." The appearance of a strong and pungent odor was one of my best indicators that it was time for me to go to the bathroom because in the absence of gravity, normal earthbound indicators did not always suffice.

Regarding convection, the ISS must use forced-air convection in the absence of gravity. If we don't do that, it is possible for CO_2 to hang precipitously and invisibly in front of unsuspecting astronauts, waiting to be pulled back into their bodies as they work, sleep, or relax. This is not a good thing.

Living on the ISS, my gas could be quite stinky. Ask a few of my fellow crew members. It could be so much so that one of my spacewalking partners—who shall remain nameless—often gave me clear verbal indications that my gas was aromatic and not in a good way. It was after repeated verbal jousting that he and I, alone in the quiet confines of the space shuttle's middeck, noted the large air hose positioned above the area leading to the shuttle-ISS docking tunnel. That air hose sent shuttle-conditioned air directly into the ISS, helping to create the forced convection. Hoping to keep aromas in the middeck at a more tolerable level, I utilized my weightlessness to lift myself to the ceiling, pointed my rear end toward the hose, and "fired away." In theory, I was sending the smelly air to the ISS, helping to maintain the middeck's air at a more aromatically pleasant composition.

Yes, I know this might seem a bit juvenile, but having read my other posts, you know this is not above me. We were just trying to have some fun during slow times on orbit. If that isn't enough, we started referring to these events as "sending emails" to the ISS. For those flatulent episodes that were noisier than usual, exhibiting a bit of machine-gun flair, we referred to them as "emails with attachments." Potty humor—do we boys ever really outgrow it? Not me.

Question: If you're on the ISS, what happens if you sneeze and fart at the same time? Do you spiral out of control?

Answer: Ah, another classic question. Contrary to popular belief (and I understand the theory has been tested, Garrett Reisman), farting in space does not really provide enough propulsion (in my case anyway) to propel a person of my mass anywhere. But it can stink up a module.

Sneezing in space is even less of a propulsion system. It works the same way as it does on Earth, unless you sneeze in your space helmet. Then you have to hope it doesn't get all over everything and does not spray your helmet visor! If I did sneeze, I quickly turned my head to one side and then wiped my nose on the piece of foam affixed to the bottom of my helmet, affectionately known as the "Dolly Parton" Valsalva device. Can you guess its shape?

Question: Has anyone actually farted in a space suit?

Answer: I sure did. Many times, I'm guessing, both inside *and* outside of a space suit. The latter was undoubtedly much to the chagrin of my crewmates. After all, I do love onions!

Question: Why do astronauts need to shower in space, and how important is hygiene aboard the International Space Station?

Answer: First of all, to be totally correct, we do not shower in space. The capability for using an actual shower has not been available since the U.S. Skylab space station days. To remain squeaky clean on the ISS, we must make use of towel baths. (The shuttle was a slightly different story, but towel baths were still required.)

Personally, my cleanup routine was one of the highlights of my day. After a good solid daily workout, I truly enjoyed my time in the solitude of a thirty-minute cleansing period. It was psychologically refreshing, making me feel clean and invigorated for the events that lay ahead.

During my 152 days on the ISS, the Expedition 15 crew had an area in the Russian functional cargo block module that we dubbed the hygiene station. This hygiene station was not universally acknowledged as being on the ISS. Much like Lord Voldemort of Harry Potter fame, it was not to be mentioned, as the ground control team would have tried to remove it since it wasn't exactly "per regulations." However, it was very convenient, easy to use, and functional for our purposes.

In this area, near its forward hatch connecting it to the pressurized mating adapter module of the U.S. segment, a small space was cleared as a dedicated area for cleanup. Behind an overhead panel was a water bag connected to a hose and pump combination of Russian origin with a spigot on the end of the hose. In this area, we could do all the things we needed to keep ourselves tidy and inoffensive to our crewmates and visitors: bathe, shave, wash our hair, brush our teeth, and so on.

I cleaned up each and every day. By asking the Mission Control Center team (and, yes, they do control nearly every minute of every day) to schedule my exercise periods first thing in the morning after breakfast, I could wake up, don my exercise clothes, enjoy a quick breakfast, and then head to the "gym." Once I completed my weight lifting (or resistance training, since there is no gravitational force of consequence) and aerobic sessions—either on the bicycle or the treadmill as we alternated on a daily basis—I headed to the hygiene station for a much needed cleanup.

The process was straightforward for me, and I started by cleaning my body. Stripped down to utter nakedness (when it was only men on board), I used one of the Russian prepackaged wet towels and began to wipe off my entire self. Contained in a small (about three-by-three-inch) package, this towel was premoistened with both water and a small amount of disinfectant. Personally, I stretched the same towel's use for three straight days, adding water from our hygiene station to moisten it again on days 2 and 3. Although some disinfectant remained, on days 2 and 3 injecting additional moisture into the towel made it softer and more pliable since its initial use.

I continued my cleanup ritual by washing my hair. Still almost naked, having re-donned my white boxer shorts, I grabbed a handful of U.S. no-rinse shampoo and a touch of water from the water hose. (I rinsed anyway 'cause leaving that nasty stuff on top of my follicly challenged cranium would *surely* have made my head itch!) Moving my hands gently over my head—zero gravity would cause the water to fly off in every direction if I was too vigorous—I could clean my nearly hairless head pretty easily, getting a good scrub and massage from my strong fingertips. I followed with a good rinse, using more water, to remove any remaining shampoo. After a methodical drying, using a rough and quite ugly, orange-striped Russian dry towel, it was time to move on to shaving.

For shaving , I used Edge-brand shaving gel and a Gillette Mach3 razor for an extremely close and refreshing shave. (I didn't get a close-enough shave with the available shuttle-era electric razor.) I simply squirted some water on my face and strategically placed several small dabs of the shaving gel around my cheeks, chin, and neck area. Performed just as I do on Earth, the toughest part about shaving for me was how to clean the razor blades after a couple of swipes. With no running water on board, I had to wipe the blades with a white Russian-brand gauze pad (similar to one you might use to bandage a child's scraped knee) that I threw into the "wet" trash bag afterward. Anything that contains moisture and has the possibility of smelling over time, such as empty food packets or my shaving-refuse gauze pad, goes in the wet trash. Dry trash would be something like a sheet of notebook paper or an index card.

I used a standard cotton towel to dry my head and face, and after every use, I attached the towel to the ISS wall, so our environmental control system could suck up the remaining moisture therein and convert it to drinking water later. I tucked each corner under a bungee strap and pooched the towel from the wall, leaving an air pocket behind it.

The act of teeth brushing was pretty normal, with the exception of having to either spit into a towel or swallow the remains. I swallowed it all, creating a nice after-dinner breath mint and at a minimal risk to my overall health. (Some folks out there in the social media world freaked when I told them I swallowed my toothpaste.)

Now, reasonably clean again, I would put on fresh clothes. *Truly* fresh clothes (right out of the package) were the best, but usually I had to put on clothes I had already been wearing, since each clothing item had to be worn a specific number of days before reaching its disposal limit. At the end of my personal cleanup session, I added a small touch of the Russian aftershave that was so generously provided for me in my Russian-made Dopp kit. Not too powerful but a manly scent, it kept me smelling good the entire day. Ready to face the day's activities, I'd head for the U.S. lab module, my "office" in outer space.

As well-known space expert and NASA engineer Robert Frost so eloquently states: "Six people are trapped in an enclosed volume. Hygiene is of paramount importance." I agree! Recalling my Russian crewmate Kotov floating effortlessly toward me in his usually smelly gym clothes is not one of my fondest ISS memories.

Question: Do astronauts use disposable towels, or do they wash them?

Answer: There are no laundry facilities on today's ISS. Someday a newer space station might have that capability. However, for the foreseeable future, washing clothes "will not be an option." (See how I did that? Another great space reference!)

On day 3 of my cleaning ritual, when I used my bath towel for the final time, I still had to let it air dry before folding it down the middle once and wrapping it tightly into a cylinder for disposal in our laundry bag with other used items such as underwear, shirts, socks, and so forth. These soft-stowage objects could then be used to stuff or fill holes in the Progress cargo ship, which would eventually undock and burn up in the earth's atmosphere, along with all of our dirty clothes and waste.

Question: Would an ISS module dedicated to relaxation help morale? Every image I see of the interior of the ISS makes it look like a mess, a nest of rats' tails, with stuff projecting everywhere to catch people out and confuse them. I'm sure it's clean and tidy in its way, but I wouldn't find it relaxing. If I had to live in a confined space like that, I'd appreciate a dedicated lounge area, where the walls were tidy and the furnishings were soft, with seats with straps to hold me in; a home cinema; special controllable mood lighting; and other stuff. I realize it would be impractical and expensive, but would it be welcomed by the astronauts?

Answer: This is an excellent question, and my personal answer is yes, a module dedicated to relaxation would be a help to morale and life in general in outer space. Here's why I feel that way.

The life of astronauts on board the ISS is quite remarkable, and they are kept very busy. These highly motivated individuals, usually with type A personalities, have accomplished great things in their careers. Their time in space is valuable, and they remain clearly focused for most of each and every day. Yet there were times for me when a focused relaxation capability module might have been an asset.

Typically, astronauts seek relaxation through music (listening to recordings and playing instruments), photography, reading, watching TV and movies, and playing video games. Some even played chess. During my stint on board the ISS, I did many of those things and even played my version of Trivial Pursuit with the ground control centers. (It was too easy for them; they googled everything.) But none of that truly helped those times when my focus turned to my mother, who was battling lung cancer. Her condition weighed heavily on my thoughts.

Chip Davis—a fellow Nebraskan, good friend, and world-renowned drummer and founder of Mannheim Steamroller—has been working on a technology capability he calls ambient therapy (http://www .ambiencemedical.com). Through the use of psychoacoustics and the philosophy that you "can believe you are where your ears tell you that you are," Chip has created a system using high-resolution digital sound algorithms for medical treatments ranging from pain distraction and claustrophobia to counteracting feelings of isolation and anxiety. His concept, which uses digitally recorded soundtracks of specific destinations (such as the rain forest, high desert plateaus, birds in a forest, etc.), is already in use at the Mayo Clinic and around the United States, with planned uses to aid our U.S. military veterans in their battles with posttraumatic stress disorder.

All five versions of Chip's Ambient Therapy System integrate music and natural sound algorithms with color therapy, and being in a form that is easily portable, they may be used during travel, even space travel. Chip has presented his therapy system to the chief scientists at the Johnson Space Center for inclusion as an experiment on the ISS, and NASA approved the use of a unit but has yet to fly it in space. I certainly wish I was still a flight-eligible astronaut, as I would be a strong advocate for his experimental idea. Imagine its potential—including coupling it with virtual reality—as we begin to deal with yearlong flights on the ISS, or six- to nine-month-long trips to Mars, or deployed astronauts who must cope with the death of a loved one on Earth.

Question: What is it like to sleep in microgravity?

Answer: I love to sleep. If there were an Olympic medal for sleeping, I'd be right there, fighting for the gold. My experience in the sleep department spans several assorted venues: in the deep woods on camp cots with Boy Scouts, in sleeping bags on snow-covered Russian and Canadian tundras covered only with branches and parachute canvas, on a steel-hardened bunk bed hanging inside a metal tube affixed some sixty-five feet beneath the sea. But the ultimate venue—a sleep station free from the encumbrance of gravity—was located 225 nautical miles above Earth!

As a U.S. astronaut, I learned that sleeping without gravity is both a challenge and an adventure. You may sleep in any orientation you choose. Absent gravity, the only true needs are adequate ventilation (including cooling), peace and quiet, and a few tie-downs. The bed of choice is a Russian-designed sleeping bag on the ISS, and we used bags of U.S. vintage on the shuttle. Both include a liner and zip-up front, providing adequate warmth, easy entry, and even easier exit. A final, key feature is armholes, allowing sleeping astronauts to take on the appearance of a comatose, stationary zombie as their arms float extended forward and parallel to the ground.

My first night of sleep on a shuttle was lacking, to say the least. In space for the first time, I watched fellow crewmates as they jockeyed for prime positions on the shuttle's middeck. A true rookie, I waited to see what was left. Tied to *Atlantis*'s forward wall housing middeck lockers, I slept vertically with my head toward the ceiling in hopes of somehow fooling myself into thinking I was back on Earth. Clad in boxer shorts and a short-sleeve T-shirt, I donned a Lone Ranger–like eye mask (sans eyeholes) to help ward off any light sneaking in systematically from the sunrises and sunsets offered via orbital mechanics every ninety minutes. I awoke the next day tired after a very restless night.

The ISS was different. My bunk was in the U.S. sleep station. It had everything I needed. Barely the size of a phone booth (young kids, do a Google search!), it was freezing cold given the great air-conditioning, nearly totally dark—and with my eye-covering mask, completely so—and, for the most part, peacefully quiet. Yet it would take nearly a week for me to finally achieve enough hours of sleep I could deem reasonable.

My evening ritual was always the same. After brushing my teeth and removing my clothing—everything except a pair of white boxer shorts—I flew into my open-zippered sleeping bag. Maintaining my shuttle-born heads-up vertical orientation, I stuffed my arms through the ergonomically placed holes, closed the zipper about halfway, and grabbed my paperback copy of Clive Cussler's *Sahara*. Exhausted from the day's combination of exercise, equipment maintenance, science experiments, and an ever-increasing CO_2 buildup in the station, sleep came easily, making my eyelids heavy and ready to close.

As I reached a repeatable rhythm of sorts in the sleep department, I would relearn that the brain is an incredible thing. With several nights of decent sleep under my belt (I averaged about seven hours and twenty minutes per night, or much better than I do on Earth!), I discovered that during those initial stages of falling asleep, my brain was telling my body I was lying on a mattress. Slowly drifting off, I felt pressure points up and down the entire length of my spine as if I were back in Houston and lying horizontally on a firm, supporting mattress. Amazing!

I was not a napper in space although I am on Earth. As I recall, I took a "power nap" only once.

My sleep time in space was, minus the lack of gravity, typical. I was frequently awakened by minor annoyances, just like on Earth. Unplanned fire alarms and the need for a timely trip to the bathroom were examples of random sleep disrupters that are similar to those experienced in everyday life on our home planet. Was it difficult to sleep in space? Not a chance! And I wouldn't change a minute of it.

I think it's time for a nap. Sweet dreams!

Question: What was your best experience in space?

Answer: Wow, that's kind of a cool question, and it's a bit of a tough one to answer!

I had a *ton* of great experiences in space: first arriving on-orbit, living and working on the International Space Station for longer than five months, performing six space walks, welcoming "visitors from another planet" (STS-118, STS-120, Soyuz TMA-11), and playing in near-zero gravity. Each and every one of them were out of this world. So what was my *best* experience, you ask?

After careful thought and further review, I think my best experience in space was carrying out a successful mission. Along with my ISS crewmates, Yurchikin and Kotov, it was undoubtedly our combined efforts in bringing about total mission success. We were rookies in outer space—all three of us—and some key players in the Astronaut Office didn't let us forget it. They questioned our crew readiness; they questioned our spacewalking and our language abilities. Lord only knows what else they felt we were not ready for.

But we did it! We did everything they asked of us and then some, and we did it without making mistakes. I would call that true mission success, and I would definitely call it my best experience in space.

Question: How does it feel to reenter the atmosphere?

Answer: It's important to note that both times I reentered our Earth's atmosphere, I was riding in the space shuttle. I have never participated in a Russian Soyuz reentry. While many points are similar, there are a couple of key differences.

The Soyuz and shuttle trajectories—how they're calculated and executed, as well as their landing point target/selection—are essentially the result of orbital mechanics and the systems' capabilities of the spaceships. However, once the vehicles are captured by the atmosphere and head toward landing, things vary slightly.

Both plummet into the atmosphere at an "angle of attack." This helps to slow them down and exposes their protective heat shields to the brunt of the friction generated as they bore deeper into the atmosphere. The shuttle pitches its nose up like an airplane, because its thermal protection tiles are on its belly. Since the Soyuz is a capsule (think gumdrop shape), its fatter and blunter bottom leads the way, and its heat shield ablator protects the ship from the high temperatures. Both vehicles are capable of moving gently against the atmosphere by rolling to the right and the left, helping to further reduce their speed and more accurately target their landing site. The shuttle's aerodynamics continue to aid its airplane-like return, but the Soyuz must deploy parachutes to reduce its speed enough for a safe landing.

The shuttle requires a runway to land, but a Soyuz can land anywhere on the Earth: desert, mountains, forest, oceans, swamps. The *desired* Soyuz landing spot is on land, somewhere in the Russian Federation. The final descent is controlled by parachutes, and small thrusters fire about ten feet above the ground to help slow the capsule down just before impact. Regardless of those thrusters, American astronauts have described the landing as "a controlled car crash," or quite dynamic and impactful. We are trained to keep our mouths closed and heads firmly against the back of our seat to avoid any car crash–like injuries in the event of a rough touchdown.

The shuttle's reentry was pretty benign, again much similar to that of an airliner. By that I mean there wasn't a lot of shaking, turbulence, and the like. To me it seemed pretty calm for the most part. But with respect to the shuttle, it's not so much the actual reentry and landing as the day's preparation to come home that dictates how your day will go. Specifically, the activity called fluid loading—that is, replenishing the liquids shed from your body while living in microgravity—can have some interesting consequences as some highlights from *The Ordinary Spaceman* reveal:

> Being a rookie astronaut, my knowledge of the fluid loading process was purely theoretical. My choice of beverages to load on landing day was based on an earthbound set of taste buds and included hot chicken broth (two bags—a big mistake), followed by purple grape drink and ending with tropical punch. The fruit drinks were artificially sweet-ened, as sugar diminishes the body's ability to retain liquids, thereby lessening the benefit of fluid loading. Since my size and weight dic-tated that I needed sixty-four ounces of fluid over a two-hour period, I followed up my tasty combination of liquids with plain old iodinated water and the remaining mandatory salt tablets. The process began with a call from our commander to "initiate fluid loading." I was not prepared for the ramifications of that simple directive. . . .
>
> When [Commander Pam Melroy] Pambo called for us to load our second bag, I wasn't even half finished with the first one and I still had purple grape drink, tropical punch, and water to get down. Fluid loading was not as easy as I'd thought it would be.
>
> With our preparatory tasks completed . . . our focus turned to the upcoming [orbital maneuvering system] OMS burn.
>
> Critical to entry if completed successfully, it would drastically reduce our orbit's altitude by slowing our orbital velocity. With the shuttle moving more slowly, the effect of gravity becomes more pronounced, pulling us toward Earth, which is the desired effect. The timeline

would then begin to move at a breakneck speed, giving us less than sixty minutes before touching down. . . .

Once . . . strapped . . . in prior to deorbit, there was nothing for me to do but relax and take in the experience and the final activities performed by my crewmates."

Question: After floating around in the ISS for many days, how does it feel to use your feet and walk when you get back to Earth?

Answer: I spent 152 days living and working on the ISS, constantly under the influence of minimal gravitational forces. Microgravity, a wonderful boon for astronauts while in space, can create the potential for problems when those same astronauts return to Earth and face the full brunt of its gravitational tug.

When I returned to Earth, I felt extremely heavy—two hundred pounds heavy, as a matter of fact. I shuffled when I walked for the first couple of hours, and I had a huge urge to "rotate right"—right into the wall. The fluids in my brain were still "spinning" as if I had stayed in low-earth orbit. It took a while before my personal gray matter was fully back on Earth with me. But when it was, I recovered quickly, no doubt due to my dedication in doing my daily two and half hours of exercise on the ISS.

I recovered so quickly that the next morning—after a solid night's sleep—I was able to walk erect without the desire to turn into structures and without the aid of someone holding me up. (I let my two kids *think* I needed them holding me, as it felt so awesome and "dad-like.") I could stand, sit, and walk easily. But sitting was painful because, after all, I hadn't done it in five months. Truthfully, I felt pretty darned good.

Many astronauts have different stories about their returns and their feet or walking. Some had tender feet, needing special footwear (e.g., Crocs) to get around. Others felt pain from the lack of muscle use. It took many a few days before they could begin their rehabilitation, which focused heavily on balance and getting around normally on one's feet. I had none of that. I rode the stationary bike for fifteen minutes the first day home and then initiated my full regimen of rehab with NASA physical trainer Mark Guilliams the next day. I was fully back to normal in only three weeks' time. (I have been told six weeks is typical for a full recovery.)

Question: What is the control stick between the legs of the Soyuz cosmonauts and astronauts-passengers? It looks like a typical stick found in airplanes, but airplanes, which also have rudders and ailerons, take more than just the stick to operate. If the ship can move with six degrees of freedom (referring to the directions a rigid body can move in three-dimensional space: back, forward, up, down, right, and left), then how does a stick with only two degrees of freedom control it? I know it would be for emergency use only, but I fail to see how it would work at all.

Answer: Honestly? It does absolutely nothing at all, unless you consider it a "mental pacifier." I was told in my training by our Russian Soyuz instructors that most Soyuz fliers were pilot types—especially in the early days (much like NASA's)—and the control stick was put there to give them something to do with their hands. It serves as a kind of mental release. The Soyuz, in most operational modes, is highly automated, giving the spacecraft's "drivers" little to do except monitor the vehicle's systems.

During the short Soyuz flight we executed on Expedition 15—we moved the Soyuz from one docking port to another to allow for another approaching Soyuz (Expedition 16)—I'm not sure anyone ever touched the stick except to move it out of the way.

6

Space Celebrity and Miscellany

Question: Why do you often write "keep lookin' up" at the end of your answers to questions?

Answer: As an author, I look for ways to stand out. With the advent of a powerful internet that leads to self-publishing, social media platforms, and blogs, a tremendous number of authors are out there. But I'm selfish. I want folks to look for me and my works. I want them to seek me out because they want to hear what I have to say. When I joined Quora many years ago, for example, I began responding to questions for two very simple reasons: I love to write, and I want to share the exploits of an amazing astronaut career.

It wasn't long after my writing debut that Jack Dunn, a dear friend from the University of Nebraska, gave me an idea. Jack worked for many years at the university, serving as the curator of the Mueller Planetarium in the Morrill Hall/State Museum. Prior to his well-deserved retirement, he was instrumental in the planetarium's upgrade that would allow incredible space laser shows to be projected in the dome theater. He mesmerized audiences by combining dazzling education and entertainment effects in a new age, exhilarating sensory experience.

Jack and I communicated often and developed a solid friendship. With Jack in Nebraska and me in Texas, south of Houston, we corresponded for years via email. At the end of each note he sent, Jack's love of the heavens came to the fore. He signed each of them with the phrase "keep lookin' up." That phrase always made me smile.

So when I began spending time on social media platforms, I looked for added ways to make my posts more unique and fun. I asked Jack if I could use his trademark phrase as a sign-off for each of my answers, similar to how he ended his emails. Being the honorable friend and selfless educator he is, Jack readily agreed. You see, he knows that anything—no matter how small—capable of engaging folks to cast their eyes to the heavens is a solid investment in education. From that moment on, I have endeavored to end each of my Quora answers with the phrase "keep lookin' up." It has become somewhat of a trademark for my answers, with people calling me out when they fail to see it.

Thanks, Jack, and keep lookin' up, everyone!

Question: What do astronauts think of *The Martian* (2015) movie?

Answer: My friends on Quora helped me learn how to become a player within the world of movie reviews with regard to this answer. I wanted to post my previously written review of *The Martian* on Quora; however, I wasn't quite sure how to do it unless someone submitted a question to me about it. Since that hadn't happened, I piggybacked on a question submitted asking about a review of the movie *Interstellar*. Hey, as of this writing, I'm fifty-eight years old, okay? Social media and computer apps are not my forte.

As for *The Martian*, I thought it was pretty entertaining, and the news website Quartz *did* ask me to write a review. It is re-posted here (from *Quartz* on September 29, 2015).

Astronaut disclaimer: I have never written a movie review until now. Of course, prior to 1998 I had never done any astronaut stuff either, and that turned out pretty well. So here goes nothing.

Recently, I was given a great opportunity. Along with an entourage of NASA scientists, engineers, management, and administrative/support staff, my colleagues and I entered a packed movie theater united in pursuit of the day's mission. Per our timeline, we purchased popcorn and drinks, performed last minute checks of our social media sites, and then sat back to enjoy a preview of *The Martian*, courtesy of 20th Century Fox.

Highly anticipated by those of us who consider ourselves space geeks, the movie did not disappoint. Oh sure, there were some technical "that's not quite right" glitches (mostly met with chuckles throughout the theater) but overall, I was quite entertained by the sci-fi tale of courage and ingenuity on the Red Planet.

Director Sir Ridley Scott (who, by the way, did a lovely video thank you/tribute to NASA for its cooperation in making the film) succeeds in bringing author Andy Weir's 2011 novel to the screen with help from a well-conceived screenplay adaptation from Drew Goddard. (Nice space name!)

Having already read and enjoyed Weir's excellent adventure, I was pleasantly surprised with the effective presentation of the novel on screen. Seeing beautiful Martian vistas, punctuated by mountainous terrain in variegated hues of orange, made it seem as if humans were already living there. The use of high-altitude and digitally accurate perspectives of the Martian surface pulled at my heart strings. And I loved that Andy Weir developed a relationship with NASA after publishing the novel, leading the push to involve the space program directly with Scott. The resulting emphasis on science provides an enjoyable balance between the film's considerable entertainment value and its educational, inspirational, and technological references.

What truly stood out in my mind, however, was not the beautiful cinematography. Nor was it the enticing musical score, or superb acting led by an appropriately irreverent Matt Damon, playing the stranded astronaut Mark Watney. Rather for me, the highlight was the film's refreshing and inspiring depiction of NASA. I'm not talking about physical depictions mind you (the Vertical Assembly Building does not reside at the Johnson Space Center) but instead the film's sense of an ever-present drive on the part of NASA employees to pull together to win the day, even in the midst of seemingly insurmountable odds. Just as I witnessed so often throughout my own 30-year NASA career, a team of ordinary, caring people with little regard to their personal needs put in just a little bit extra, to do something extraordinary.

I felt an immense sense of pride as I witnessed the fictional yet plausible teamwork and collaboration between NASA's unmanned robotic programs—housed at Pasadena, California's Jet Propulsion Laboratory—and its human-mission teams at the more recognizable Mission Control Center in Houston.

NASA, more recently depicted through the lens of the mundane, reduced to a series of news articles about budgetary standoffs and political squabbles rises above it all in this fictional thriller. In marked contrast to news media's reality, *The Martian* paints what I believe is a much truer picture of our nation's space program, one that requires the collaborative efforts of all teams in order to successfully explore whatever it is that's potentially "out there."

To that point, I reveled in the scenes of international cooperation. Not between today's actual international partners of the space station (15 strong), but with China, the newcomer in the race for space. A science fiction survival and rescue story in which one of the US's current adversaries plays a key role in the mission's success? What a tantalizing and hopeful vision for the future!

Ultimately, in spite of its technical flaws (I leave those points to scientists more technically qualified than I—that's your cue, Neil de Grasse Tyson), I liked *The Martian*. I am hopeful that many other earthlings will like it too. And maybe even a few Martians!

The moral is clear: Add extra to ordinary, and you, like Mark Watney, can be extraordinary! A perfect metaphor for our space program, and an inspirational message for anyone aspiring to push themselves to greatness.

Question: Is it possible to see fireworks from outer space?

Answer: I was up in space during the Fourth of July holiday. I planned to try and see if I could actually view fireworks from space. However, my plans were doused when I was unable to stay awake that long. Due to the ISS being on Greenwich mean time, or Zulu time, we were six hours ahead of the U.S. East Coast. So in order to even try and see fireworks, I would have had to stay awake until about 4:00 a.m. or get up early. I chose to do neither.

Question: Can astronauts aboard the ISS see the Moon from any of the windows or while doing an extravehicular activity?

19. The Moon rises above our home planet.

Answer: Why, of course! As a matter of fact, I saw a classic and beautiful moonrise during my first space walk on July 23, 2007. As I was working on the front of the ISS, near what we call the external stowage platform 2, I paused for a second to catch my bearings. I looked up as a full moon was quickly rising behind the ISS. It absolutely took my breath away.

Question: Do any astronauts or former astronauts maintain blogs?

Answer: Yes, and I would like to add my name to the mix. While I don't necessarily have a blog, per se, I do have my website, astroclay.com, and am very active on Twitter (@Astro_Clay) and Facebook (Clayton C. Anderson). I also write articles occasionally for various news outlets such as the *Huffington Post*, *Forbes*, and *Newsweek*. I would be honored if you would check them out and let me know your thoughts.

Question: What are the implications if two member countries of the ISS go to war with each other?

Answer: Honestly? I hope we never find out.

The ISS stands as a symbol of cooperation between thousands of people spanning our globe. Originally a project involving sixteen individual space-focused nations, it boasts an operational presence in space now stretching almost twenty years! With a little luck and continued Earth-based cooperation, we'll continue our space-based efforts out to the year 2024 and beyond.

The ISS is not only the world's most complex endeavor performed off the planet but it's also a symbol—a symbol of what humans can achieve together when they throw away the trappings of jealousy and hatred, of fear and intolerance. To me, it is a symbol of *all that is good in this world*. We need more symbols like that.

I have always been and always will be a huge fan of *Star Trek*. From my boyhood days in Nebraska, I looked forward to the adventures of Kirk, Spock, Bones McCoy, and Scotty as they traveled the universe and exercised "the prime directive." In that forward-thinking television program, the United Federation of Planets stood as an example of our future. How about, for now, if we just continued to expand on the concept of cooperation right here on Earth?

"Live long and prosper!"

Question: Can you cry in space?

Answer: I cried in space several times due to some very emotional circumstances. Crying is exactly the same as here on Earth, except the tears don't fall down as there is no gravity. Not a big deal at all; however, the emotions I experienced were a big deal. Some were because of personal events and some due to circumstances inflicted upon my family by folks at NASA. Check out my first book *The Ordinary Spaceman*. Should make for some interesting reading.

Question: What restaurants in the world have the best views?

Answer: Not sure if the ISS is considered a restaurant, but having dined in its cupola (shhh, we probably weren't supposed to take food in there!), I would have to say that it had one of the best views. It's "out of this world!" The restaurant's food was pretty average, but the ambience was stellar.

Question: Has anyone ever taken a Bible to space?

Answer: Yes! I had an electronic Bible with me during my long-duration mission to the ISS in 2007. It was the revised King James version.

Question: How would you build a cheap, inexpensive space suit replica for Halloween?

Answer: Already been done, but it wasn't for Halloween! Way back in 1965—for my hometown's big summer celebration (I was six years old)—we had a kiddie parade, which was nothing more than a youngster's costume contest. My mother wrapped my entire body in aluminum foil. She took one of her hat boxes, and after cutting holes for my eyes and an opening to allow me to put it over my head, she wrapped it in Reynolds Wrap aluminum foil as well. She added a pipe cleaner attached to a Styrofoam ball (so I could communicate with the aliens) that was also covered in foil, and—voilà!—I was a Mercury astronaut.

Question: What is the room in NASA where scientists make the candidates for astronauts perform physical exercises called, and what does it look like?

Answer: There is no such room that I know of. The astronauts have an exercise facility that helps to prepare them for long-duration spaceflight and the necessary rehab upon their return to Earth. These days astronaut wannabees who come to Houston for the formal interviews are invited to participate (it's really mandatory, because, you don't want to be the "hopeful" that *doesn't* participate!) in some physical exercise sessions called CrossFit. This latest trend in exercise is essentially a cross between weight lifting, stretching, and aerobics. Suffice it to say that if you want to be selected as an astronaut, you'd better be ready to do it well as astronaut veterans also attended the exercise hazing to see if you have "the right stuff."

Question: Have you ever met a fan of yours in real life?

Answer: Yeah. And I married her!

Question: Why do astronauts seem to wear up to three wristwatches?

Answer: Good question, simple answer! The usually black colored watches seen in some NASA published pictures are part of an experiment called SLEEP (sleep-awake actigraphy and light exposure during spaceflight) Long. As I recall, the experiment requires the astronaut to wear a special watch that is capable of "seeing" light and dark and of sensing movement and no movement. Over time, astronauts use a laptop computer to download the data from the watch to the ground investigators. This onboard zero-gravity data is compared with data gathered on Earth before and after the flight. The scientists can then tell how well and how long the astronauts slept during the night throughout their time in orbit. (As noted previously, I averaged about seven hours and twenty minutes over 152 days.) This data will lead to possible ways to increase and enhance an astronaut's ability to get meaningful, restful sleep while in space. For many astronauts, sleep does not come easy during a mission due to many factors like stress, excitement, adaptability, and so on. Applications to those of us on the ground are many, including ways to alter circadian rhythms that could help emergency room doctors, long-haul truck drivers, and even Mission Control Center flight controllers, all of whom must sometimes work long—and nontraditional time—shifts.

Astronauts also wear watches to simply tell the time, and many wear multiple versions so they can track the onboard, or Greenwich, time; their family's local time on Earth; and perhaps a third time of interest (e.g., that in Russia).

Oops, just checked my watch. Almost time to stop typing.

Question: Do you, Clayton C. Anderson, answer questions on your own, or do you have a media management team?

Answer: I, Clayton C. Anderson, answer questions posted on social media platforms such as Quora and anywhere else *on my own*. I do not have a media management team—I can't afford one—unlike some of my colleagues. I *love* trying to inform folks who ask questions via social media platforms, but I provide answers that are anecdotal in nature. I choose to not provide a bunch of highly technical information. You can google that stuff!

By the way, my first book, *The Ordinary Spaceman*, did *not* have a ghostwriter or a coauthor, and I am very proud of that fact.

Keep lookin' up!

7

Philosophy and Politics

Question. Regarding conspiracy theories, is the International Space Station fake?

Answer: Hmm, is the ISS fake? Is it all a conspiracy theory? Well, if it is, it's a pretty damned good one.

I'm never quite sure where people come up with this stuff: the Earth is flat, and the Moon landings were filmed on a studio back lot in Hollywood. And now you are suggesting to me that the ISS is faked. Do folks like you have nothing better to do with your time than to propagate this junk?

I lived and worked on the ISS for 151 days, eighteen hours, twenty-three minutes, and fourteen seconds. I added another 15-plus days on a second trip via the space shuttle *Discovery*. I viewed our planet from 225 nautical miles away and saw the curvature of its horizon. I sat with men who lived and worked on the Moon, learning from their experiences and, with amazed awe, listening to these heroes tell stories. I played in the free-fall microgravity environment for five months, spinning, twisting, and flying like Superman.

I think sometimes readers ask me inane questions such as this one just to see what kind of reaction they'll get from me. As usual, I aim not to disappoint. I encourage readers with such views to focus their energies on something more important than promoting ridiculousness. Maybe try running for president?

Question: Are there laws regarding the direction of and heights for satellites orbiting Earth, and if so, what agency developed and enforces these laws?

Answer: I don't know of any laws that exist to cover these types of things. However, I do know that the realm of "Space Law" is coming to the fore these days with programs cropping up in many universities (e.g., University of Nebraska–Lincoln). We must begin to answer myriad questions with regard to how we utilize space commercially now that multiple countries have shown the ability to get there.

The laws of orbital mechanics dictate many of the parameters you mention. For example, launching to the east gives an added boost from the Earth's rotation, thus saving fuel costs. Geostationary satellite orbits, which place a satellite over the same point on Earth, are necessary for many types of communication satellites. Basically, the job the spacecraft needs to do dictates where, how high, what inclination, and so on are required for its launch and eventual placement in orbit.

Question: What do the people at U.S., European, and Russian space agencies think about the success of India's Mars Orbiter Mission, particularly in light of the extremely low cost of the mission (less than one-tenth of a similar NASA mission) and the frugal innovations that made it possible? Do they think of it as just another "me-too" gimmick by a third world country to enter the big league, or can they foresee some tangible techno-commercial benefits of this endeavor?

Answer: I am not sure what other space agencies said, but I am certainly hoping that they—as I do—think it's quite awesome! We are a space-faring planet, and we all need to work together to explore and pioneer the heavens. We welcome India into the space race.

And, of course, we should all endeavor to do it together, peacefully, and in the spirit of preserving humanity. We look forward to their future successes, as well as those from our friends in China.

In 2014, Charlie Bolden, NASA administrator, made the following statement regarding India's Mars Orbiter Mission (MOM):

> We congratulate the Indian Space Research Organisation for its successful arrival at Mars with the Mars Orbiter Mission.
>
> It was an impressive engineering feat, and we welcome India to the family of nations studying another facet of the Red Planet. We look

forward to MOM adding to the knowledge the international community is gathering with the other spacecraft at Mars.

All space exploration expands the frontiers of scientific knowledge and improves life for everyone on Earth. We commend this significant milestone for India.

Question: Are any other astronauts jealous of Russian Cosmonaut Gennady Padalka's time in space?

Answer: Heck, yeah! Of course, I'm jealous of Gennady Padalka's massive quantity of time in space. He's a record holder! But it does come with a price. For me, it's too high of a price.

Having lived and worked in space for a total of 167 days—far short of Padalka's number, 878-plus days!—I have a small understanding of what is required to pull off a feat as mammoth as Padalka's.

I know Gennady but not as well as I know some other cosmonauts. I never flew with him, but I did spend some time with him in Star City, Russia (the training venue for cosmonauts), in brief NASA training exercises, and in social situations both in Russia and the United States. I know him as an extremely intelligent, very likable, and highly professional cosmonaut with a lovely family.

Through my Expedition 15 crewmates on board the ISS, I also know that for a cosmonaut flying in space the monetary rewards can be significant. My understanding was that in 2007, a cosmonaut could earn ten times his annual salary during a six-month ISS mission. And at that time, a typical cosmonaut salary was about U.S.$6,000. I have no authentic data to back up this claim, only chats at 225 nautical miles above the Earth. And it is possible that I misunderstood what was being communicated, so please don't shoot the messenger.

I see Gennady Padalka as "the cosmonaut's cosmonaut." I envision that he is an example to which many young cosmonauts will strive to emulate. I would have been honored to fly with him, and it would have been a blast. His ability to speak the English language is flawless and his personality magnetic. For the Russians to have someone like him holding the record speaks very highly of their cosmonaut selection process.

I understand that he's now retired and will never fly again. But from my perspective, I could not have done what he did. I needed to be on Earth with my family. They needed me and I needed them. To fly five missions stretching over two and a half years would not have been fair to my family. The sacrifices they made for me to be able to fly the two times I did were significant, and I do not take their sacrifices lightly. As I have said many times, for me to fly again, the tax on my family would be higher than what I am willing to pay. But that's me, and that's my value system.

Since I don't know Gennady's personal situation, I will not speculate. Perhaps the reason he flew so frequently is because he is a damn good cosmonaut. Perhaps it was because he loves spaceflight or perhaps because he loves his country. Those reasons all sound pretty good to me, and I'm just the Ordinary Spaceman.

Question: Are there any regional or zonal restrictions in the space station? Can American astronauts freely visit the Russian Operations Segment and vice versa?

Answer: Regional or zonal restrictions on the ISS do exist in my opinion, but we must be careful as to how we categorize them.

Having lived on board the ISS for five months in 2007, and then visiting for two weeks in 2010, I have firsthand experience as to how life is conducted internationally in low-earth orbit. As the lone American astronaut during most of Expedition 15 to the ISS, I had the privilege of living and working with two of the finest cosmonauts in history—Fyodor Yurchikin and Oleg Kotov. In my humble opinion, I truly believe that they will become Russian space legends one day. They are already legends in my mind.

As to zonal or regional restrictions on the ISS, there were none. Any such distinction is more from a set of unwritten rules that astronauts and cosmonauts are expected to follow. In addition, since my retirement in 2013, it is possible that things have changed. Let me explain.

Inside the ISS in 2007, Russian cosmonauts and American astronauts had free rein. We could easily, and for no good reason, travel between the two segments. At that time, there were no Japanese or European Space

Agency modules. Everything was either from the United States or Russia, save the Canadian robotic manipulator better known as Canadarm2.

It was understood, however, that even though we had trained on each other's ISS modules and the Russian's Soyuz systems, we should refrain from doing anything independently in the segment that wasn't "ours." For example, if I was in the Russian segment, I would not do any work on anything without the watchful eyes of Fyodor or Oleg or the clear direction of the control center in Moscow. This did not include going to the bathroom or preparing our food; we could all do that without worry of any restrictions, thank goodness. But I *will* tell you, that if the urine tank was full or the warning light in the Russian toilet came on, I immediately hollered at my crewmates for confirmation of what I believed was going on and what we should do about it.

We had the same type of arrangement within the U.S. Operations Segment. It was harder to police as Fyodor and Oleg had more U.S. training than any cosmonauts in history. By virtue of their knowledge and abilities, it was hard to slow them down when they were intent on making things right! Most times, however, they knew exactly what to do, and together, we got it done correctly.

Outside the ISS was a bit different. There was one rule that U.S. space-walkers were always required to uphold. If an American astronaut was to translate (move) over to the Russian Operations Segment, verbal permission made over the communications loops was required. More of a mental "thought jogger" than a true rule for danger avoidance, seeking permission forced us to acknowledge that we were heading into "new" territory, where the rules were a bit different. For example, the Russian segment has thrusters—with all the nastiness that comes with them. Avoiding the areas where the plume and its by-products might contaminate a space suit was helpful as far as our multimillion-dollar U.S. taxpayer investment was concerned.

As of 2010 the restrictions—if you can call them that—had begun to morph and were more monetary in nature. As a member of the STS-131 crew, I returned to the ISS at the time Oleg was serving as the station commander. Frank evening meal conversations with my former crewmate uncovered new restrictions he was being forced to consider and institute. He was not allowed, for example, to perform *any* tasks in the

U.S. Operations Segment unless permission (read: monetary transactions between space agencies) had been given from his control center team in Moscow. Neither of us felt this was a good trend. These days, I guess it reflects the cost of doing business in space.

Oh, and watch for new rules and regulations. After all, this is America.

Question: Does the ISS have spy equipment aboard?

Answer: The ISS does not have that type of capability to my knowledge. External and internal cameras are quite limited in what they can do. And, as the old joke goes, if NASA tells you too much, they'd have to shoot you!

Question: Will we go back to the Moon?

Answer: I most certainly hope so! I believe that the Moon—and a lunar colony—is the key to helping us truly figure out how we can head to and successfully live on Mars. When the movie *The Martian* hit screens across America, I wrote an article for the Huffington Post titled "Fly Me to the Moon!" That article focused on my belief that going back to the Moon should be one of NASA's priorities before tackling a journey to Mars.

You see, while most know what we want to do on Mars, we just aren't sure yet *how* to do it. And what better place to enable and test the development of all this needed technology than the Moon? A mere three day's journey away, the travel route is on known and proven trajectories. Communications delays are minimal, and we are close enough to correct process and design flaws while maintaining a needed rescue capability as we learn what is truly required to launch a viable—and sustainable—human expedition to Mars. Keep lookin' up and you'll see the Moon *and* Mars.

Question: From the perspective of danger, what should astronauts avoid doing?

Answer: Do you want the funny answer or the serious answer? Not sure? Well, let me give you a little of both!

There are obvious things an astronaut shouldn't do. For example, you shouldn't put your spacewalking boots on the wrong feet. Also, you should not forget to attach your safety tether to the tether point immediately outside the hatch before exiting said hatch on your space walk. You should not—upon receiving coupons for a lifetime's worth of free pizza at a speaking appearance—toss the coupons out into your audience, creating a mass "feeding frenzy" that may ultimately put the owner of the pizzeria out of business when numerous patrons show up at the same time for free pies.

You should not land your T-38 supersonic jet on a runway and then drive off of that runway into the muddy grass and then not tell anyone about it. You should not do an entire space walk with your elbow joint 180 degrees opposite of where it should be. You should not go outside to do a Russian space walk and forget to open the valve in your suit that allows you access to your emergency oxygen supply. Oh, and you also should not dive headfirst into the Neutral Buoyancy Laboratory to renew your SCUBA currency.

Many of you may now be thinking, Okay, enough with the jokes; tell us some real stuff. Well, according to my sources and they are pretty darned reliable, everything I listed in the preceding two paragraphs is *true*. It all has actually happened. (By the way, I was the one who dived headfirst into the NBL!)

So what have we learned here? Astronauts are not infallible. They make mistakes. Hopefully we learn from those mistakes, but sometimes we learn the hard way. You see, the real issue from my perspective is, what happens to those astronauts making the mistakes? In many cases the answer is absolutely nothing. They may or may not be counseled, and typically nothing more is said. However, in other cases, a "scarlet letter" is quietly placed into the unwitting astronaut's file, one that will follow him or her throughout the remainder of his or her career. In fact, that scarlet letter may cost the astronaut an opportunity of ever flying in space again. For example, try sending a frustrated email with profanity in the title and see what happens. I did that too. (See chapter 13, "Dark Days of Summer," in *The Ordinary Spaceman*.)

In short, as an astronaut you must realize that everyone is watching you and all the time. Always follow the mantra of "keep your head down and keep coloring." Yet eventually, we're all going to goof up something, and as an astronaut it's hard to know when and if your actions will carry severe—or no—consequences.

The so-called bar is a variable that is set high for some and quite low for others. In Canadian astronaut colonel Chris Hadfield's book *An Astronaut's Guide to Life on Earth*, he relates ways he thinks being an astronaut provides life lessons for those not in the profession. He didn't tell you about this bar, yet it exists in every walk of life.

Still, being an astronaut is something I wouldn't trade for a second. It's the greatest job in the universe.

Question: Is space exploration really a waste of money?

Answer: I absolutely *hate* this question! Why? The statement is so far from the truth it absolutely disgusts me. Let me see if I can briefly explain why.

Your ability to post the question in the first place and perhaps read my answer on your smartphone in your part of our world results from

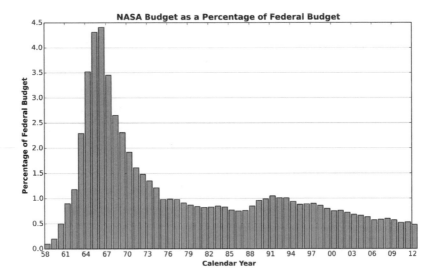

20. The support that NASA has received from the federal budget varied greatly between 1958 and 2012.

transmission satellites placed round the globe by technology developed through our space program. First developed in the 1960s, our space program—along with that of Russia, the only other country even attempting such things at the time—was in its infancy.

When someone you know grabs a portable drill and sticks in a freshly charged battery to work on a home project or repair a fence after a big storm, encourage him or her to look toward the sky, with hands extended upward in reverence, and yell, "Thank you, NASA!" This useful technology was developed during the Apollo program.

If you have a niece or a nephew fortunate enough to have invisible braces to help straighten her or his teeth but not mar that beautiful, cherubic smile, you should shake your head in awe and prayerfully say, "Thank you, NASA!"

If your grandma Martha has had a hip replacement and that hip joint is covered with a light coating of metallic gold, allowing her to walk easier than ever before, you should lift your hands to the heavens and yell, "Thank you, NASA!" Then tell Grandma you love her.

During the Apollo program, the United States spent about 1.75 cents per tax dollar on space. From about 1983 until 2013, the value hovered around 0.70 cents per tax dollar. Today that value is less than 0.50 cents per tax dollar. Depending on your source, the payback cited to all humans on Earth (not just Americans) was approximately 7 to 1. Some argue it was a high as 20 to 1. Who today would not take that kind of return on an investment? Very few, I'm guessing. Our problem is that we want instant gratification from any investment we make. We (Americans, anyway) are not willing to wait for the payback over time. If your DVR breaks, what do you do? Most go purchase a new one, if it is within their means.

Do you need even more information for me to convince you to change your tune and agree that space exploration is an investment and *not* a waste of taxpayer dollars? Just go to NASA's web page "Benefits to You" (https://www.nasa.gov/topics/benefits/index.html) and start reading. You'll be surprised at what you might find. And when you've done a little more research, cast your eyes to the skies and, following my advice to keep lookin' up, take a few seconds to yell, "Thank you, NASA!" You'll be glad you did.

Question: Is it childish to want to be an astronaut?

Answer: Heck, no, it's not childish to want to be an astronaut! As a matter of fact, I'm guessing most kids were like me and dreamed of becoming a space flier. I started when I was nine years old.

If you get a chance to read my book *The Ordinary Spaceman: From Boyhood Dreams to Astronaut*, you'll learn much about my early childhood dreams and what spurred my desire for this out-of-this-world job.

The childish dreams of little ones form the future! Whether it is the dream of becoming a doctor, lawyer, teacher, professional athlete, or astronaut, it is imperative that we encourage kids to dream big and "reach for the stars." Then, when they *do* have big dreams, we must encourage them and help them find ways to achieve those visions, period!

I tell a story at speaking engagements about how—while in high school—I wanted to go to the University of California in Los Angeles and play for legendary basketball coach John Wooden. In discussing this intent with my parents (by the way, my ACT score was 29), Dad remarked, "Isn't UCLA in California?" I replied, "Yes, Dad, it is." He followed with, "Isn't California an expensive place to live?" Again, I said, "Yes, Dad, it is." Finally, after a few seconds of thought, he ended his share of the conversation with, "Well, then, I don't think you ought to go to UCLA." And he walked out of the room. At that point, my mom placed a gentle hand on my shoulder and leaned in to whisper softly in my ear, "If that is your dream, we will find a way. . . . We will find a way."

While I did not continue on to UCLA and play basketball, I *was* still able to achieve the dream that sat at the top of my list since the age of nine, and I became an astronaut for the United States. But I couldn't have done it without my family, my friends, my teachers and coaches, my professors, my community, and my church family.

And I couldn't have done it without being a little bit childish.

Normal Spaceflight Operations

8

Question: How do you designate direction—that is, stern, bow, port, and starboard—in a space station?

Answer: To help rookie astronauts—those who journey to the International Space Station for the very first time—adapt more readily, NASA adopts conventions that make "earthly" sense. For example, while living on the ISS, you will be trained to use the terms "forward," "aft" (back), "port" (left), and "starboard" (right). We also use "deck" and "overhead" to refer to the floor and ceiling, respectively. This way we can easily—and much more clearly—communicate with our good friends on the ground in the Mission Control Center. The location coding mnemonics include dividing modules into smaller pieces such as "deck 6" or "overhead 4," based on standard rack sizes, and allow us to be more specifically directed to locate objects we need or store those we don't.

Labels abound, including directional postings that give us idiot-proof instructions on how to find other modules that may be attached below us, above us, or to the side. Specifically, it is done primarily to aid crews experiencing emergency situations, when chaos and confusion may be more prevalent.

NASA goes a bit further in this regard, as the first trip into microgravity can be quite disorienting for some. We consciously placed lights near the ceiling and power outlets near the floor, just as it is done in conventional homes on Earth. We encourage our astronauts to live on the ISS in an Earth-based attitude early in their missions—for example, with their heads toward the ceiling and feet toward the floor. These small efforts can aid one's adaptation to this unique and hellaciously fun environment.

Once an astronaut has adapted sufficiently, in a few days to a couple of weeks, then all bets are off! It's playtime in zero-g, and our brains and

21. In the U.S. lab module, directional signs—starboard, deck, overhead, To Node 1, hatch, exit—are clearly visible behind me. So are my banners from Ashland-Greenwood High School!

bodies don't really need the cues we so carefully set up in training. But we still need to be able to locate things quickly, so the conventions of location coding remain. Look around one room in your house, one that you are quite familiar with. Now, with a friend in the other room or on the end of your phone somewhere, describe where you keep something or where that something is located. It will prove challenging to both of you, giving rise to understanding why NASA does things the way it does! It makes a whole lot of sense.

Keep lookin' up (or around the room till you find it)!

Question: How do astronauts take their shower on the ISS?

Answer: As noted in chapter 5, we simply don't shower (ewwwww, gross!); however, we do take towel baths. During my five months on board the International Space Station in 2007, Oleg, Fyodor, and I had a small spot in the functional cargo block (FGB) module set aside for keeping our selves clean and spiffy.

Of course, we know that gravity is usually required to take a shower. And in the Skylab days of the mid-1970s, that space station did have a shower but no gravity. It consisted of a plastic surround-type shower curtain that pulled up from the floor and clipped at the top of the shower structure. The astronauts put water on their bodies in a fashion similar to what we do on the ISS. The difference was getting the water off of their bodies. That required using a vacuum cleaner to pull it off of them and transfer it into a tank.

Perhaps in the future, technology will advance such that we will once again have an actual shower to use in space. On the Moon or Mars, where there is one sixth and one-third the gravity of Earth, respectively, a more traditional shower may work nicely.

Question: Do astronauts take multivitamins? If so are they catered individually to each astronaut's needs, or are they a one-size-fits-all pill? If not unique to each astronaut, are they a name-brand supplement or a NASA-made concoction?

Answer: I took multivitamins every day. The brand was my choice—and shall remain nameless here until I can get a big, multimillion-dollar endorsement deal with the vitamin company—as was the taking of the vitamin. Astronauts take some meds when necessary—perhaps ibuprofen before or after a space walk—or when required for science experiments, but many refuse any medications for personal reasons. Important meds for me were my multivitamin and vitamin D. With no daily sunlight hitting my body as happens here on Earth, my bones needed some help. Help came in the forms of exercising two and a half hours a day and, for me, taking a vitamin D tablet every morning.

The vitamin D tablets came packed with our food rations, taped to the inside of the lid of the food container. It was evident to me some astronauts preceding me on board the ISS weren't taking their vitamin D regularly, as there was a huge glut of pills when I got there. The way I figured, it was my first and maybe only time in space, so I wanted to maximize my performance in orbit and when I returned to the ground. Vitamin D seemed like a no-brainer to me, with no negative impacts but a big upside for my body.

Keep taking your meds. Do you hear me, Grandma Mary?

Question: Can you drink hot drinks in space like coffee, tea, or hot chocolate? I was wondering how it would physically work. I can't imagine drinking a hot drink out of a plastic bag.

Answer: Yes, you most certainly can drink hot drinks while living and working on board the ISS. During my five-month increment in 2007, we had a system in the Russian segment that allowed us to add either very, very hot or tepid water to any of our drink bags. We sipped the drinks through a clever plastic straw mechanism that could be locked after every sip to keep the liquids from escaping in zero gravity.

The very, very hot water worked well for coffee, hot tea, Russian soups, and hot chocolate. I used tepid water for everything else. The hot water was quite hot, so much so that when filling a shiny silver U.S. drink bag or a clear Russian packet of soup, oftentimes it would be too hot to hold with your hands for the initial couple of minutes after filling it from the machine.

Even from a bag, I loved my hot coffee—Hawaiian Kona coffee with cream and sugar—nearly every morning while living on the ISS. I would usually enjoy two bags, each filled with about twelve to fourteen ounces or so (I always added more water than the instructions called for). I would fly from the Russian segment to my daily home in the U.S. laboratory module, with the two plastic straws clenched between my teeth and the bags of coffee flapping in the breeze. Upon arrival at my work station, I used Velcro to attach one bag and straw to the wall as I began to enjoy the other. While there were no steaming hot vapors arising from a solid Nebraska Cornhusker mug and no wonderful aroma ("like Folger's in your cup"), it was still a treat I looked forward to each and every morning.

On the ISS today, two such machines produce hot water. The original one is in the Russian segment, and a newer version of the Russian unit is in the U.S. lab. There is also an espresso machine, courtesy of the European Space Agency. With six crew persons on board regularly, I expect the machines get a big morning workout prepping everyone's coffee. Maybe Keurig or Mr. Coffee will come up with a space-flyable unit.

Question: What is the time line for upcoming missions for the Space Launch System (SLS)? Will NASA's launch strategy change given its latest go-ahead?

Answer: I respectfully suggest you go to NASA's website and search for details on the SLS. A good place to start is here: https://www.nasa.gov /exploration/systems/sls/overview.html. They may be located there, but it's quite early for anyone to expect published mission time lines.

In my thirty-year career as both an engineer and an astronaut at NASA's Johnson Space Center, it was quite rare that a major project was executed by its planned and publicly announced launch date. Be wary of the schedules you see for large programs such as the SLS, Orion, and so on. We may get close, but delays due to funding problems and technical issues are quite likely.

Now, regarding launch times of actual missions in the queue, I was extremely fortunate in that each and every launch associated with my two missions into space happened exactly as scheduled. The shuttle (STS-117, 118, 120, and 131), the Soyuz (TMA-10), and the Progress launches were all carried out exactly per the plan. Especially noteworthy was when I was looking for my ride home. I didn't want that one to have any problems.

Question: Can you describe what is it like to be in a rocket or shuttle launched into orbit? I can only begin to imagine the intensity of actually being in a rocket that is launched into space. Can you describe the emotions, physical sensations, and thoughts during your launch? I have only a very vague idea of what it could be like from that fun virtual reality shuttle ride at the Kennedy Space Center in Florida.

Answer: I certainly can describe what it is like to be in a space shuttle rocket launched into orbit. I have done it twice successfully, but I've never launched in a Russian Soyuz capsule perched high atop a Soyuz rocket. I'm not going to describe my experience here, since I did in detail in my book, *The Ordinary Spaceman*. In chapter 12, "Sixty-Two and Counting," I describe the emotions, physical sensations, inner thoughts, and more. It's *way* more fun than any virtual reality shuttle ride at the Kennedy Space Center! Trust me on that one.

Question: How do the ISS astronauts know where they are over the Earth?

Answer: For new ISS astronauts like me, it can be a challenge to figure out where you are above the Earth. Unlike junior high school, when doing world geography map drawing and similar lessons in map reading, the Earth doesn't have the lines drawn between the countries. Capital cities are not printed with a big star next to their names, more populated cities are not labeled in larger or bolder fonts, and rivers don't have that nice italicized name scribbled next to the wiggly lines!

So we have to rely on our own personal understanding (in my case, that's not too good) or use the tools provided for us. During my stay of 152 days, I loved to use our computer tool called World Map. When given the current ISS state vector provided by the ground control team, telling us where we were in space, the computer program could calculate our trajectory and draw our position over the map of the world. The program worked in real time, meaning our position over the Earth was constantly updated, giving us excellent knowledge of what (and whom!) we were flying over.

With time, we were able to recognize many of the landscapes and features that designated a given region. Many we already knew—for example, the Great Lakes and the Sinai Peninsula—but oftentimes I set a timer to go off when I would be over an area of particular interest I was less familiar with, such as Patagonia or islands of the South Pacific. We were also given time tags by the ground for what were called Crew Earth Observations of specific targets that the ground wanted us to capture in photographs or video images.

Question: How is New Year's Eve celebrated in space?

Answer: I cannot answer your question directly. You see, I lived in space from June 8 to November 7, 2007. According to the Julian calendar, I was obviously not living off the planet Earth during the celebration of a new year. If I had been there, then we could have celebrated several times, with multiple start times based on our sixteen sunrises/sunsets per day.

With all that being said, I can give you my insight with regard to the holidays I did celebrate during my time in space. Only three holidays come to mind: Father's Day, the Fourth of July, and Halloween.

Just days after our late afternoon (Kennedy Space Center time) launch on June 8, 2007, Father's Day approached. While the station and shuttle crews were working hard to accomplish all of the mission objectives, I decided that acknowledging Father's Day and my family might be a good thing to do. I did not want to affect the crew's schedules, nor did I want to usurp the authority of the ISS commander Fyodor Yurchikin or U.S. astronaut Suni Williams, as I was the "new guy in town." So with a little help from the ground control team and an uploaded musical file of Bob Carlisle's version of the song "Butterfly Kisses," I picked up the ISS microphone on Sunday, June 21, in the early afternoon (by central daylight time) and called down, "Houston, ISS on Space-to-Ground 2. I have a tribute I'd like to play for all of the crew's families in honor of Father's Day." As I punched the laptop's Enter key to enable the file to play, I keyed the mic and held it near the laptop speakers, while my crewmates paused briefly to see what I was up to.

As the song ended, I encouraged fathers everywhere to hug their children—especially their little girls—and wished them all a happy Father's Day.

The Fourth of July, the birthday of the United States, was the next holiday I experienced. For this special day on orbit, I had an inspiration for what I should do. Thinking back to my childhood, I recalled how television stations would end their day of programming with a video of the "red, white, and blue" majestically waving in a clear blue sky while playing our national anthem. Knowing astronauts usually sign off to the MCC prior to going to bed and using this image as my catalyst, I asked the Mission Control Team—specifically the Orbit 2 flight director Ginger Kerrick—if it would be okay for us to end the day with a playing of Francis Scott Key's work in honor of the Fourth of July. (According to Wikipedia, Key, a thirty-five-year-old lawyer and amateur poet, wrote the poem "Defence of Fort M' Henry" in 1814 after witnessing the British ships of the Royal Navy bombard Fort McHenry in the Chesapeake Bay during the War of 1812. The poem was then set to the tune of "To Anacreon in Heaven" ["The Anacreontic Song"], a popular British song written by John Stafford Smith for the Anacreontic Society, a men's social club in London.)

With Kerrick's agreement, I prepared for the moment by adjusting the lab module's video camera to view the entire module's center and removing a Velcro-backed U.S. flag patch from its usual perch just above the door on my sleep station.

At the moment of sign-off, I affixed my toes to a floor-based handrail, faced the camera while steadying myself with the U.S. flag patch in my left hand, and covered my heart with my right hand. For the entire song, I—the only American in outer space that day—floated silently in space, hand over my heart, symbolizing my love of country. As the song finished, I held my pose and then silently floated into my sleep station.

Halloween was a bit more festive for me. Having received a special package from home—delivered to me by STS-120 mission commander Pam "Pambo" Melroy—I donned my costume for the entire day of Halloween, October 31, 2007. Floating throughout the ISS as Count Dracula, I performed all of my mission's daily tasks with my black cape floating weightlessly from my shoulders. Scary? Maybe not. But I had fun, and my picture made the *Houston Chronicle* newspaper.

May God continue to bless America.

Question: How do astronauts get their hair cut while in the International Space Station?

Answer: While I can't post a video here, I do have a great one. Look for it on Facebook (www.facebook.com/astroclay), Twitter (@astro_clay), my website www.astroclay.com, and YouTube. In any event, we use a standard hair clipper set, with a few "altered for space" attachments, that is connected to our vacuum cleaner to suck up all the hair. Russian cosmonauts actually receive training in how to give a good haircut. At the time of my long-duration flight, American astronauts did not. Most receive buzz cuts ("To infinity and beyond!"), so it's not too hard. It would be a bit more complicated trimming the hair of our female fliers. (Apparently during Expedition 43 astronaut Terry Virts, who flew with Italian astronaut Samantha Cristoforetti, tried to learn how to cut Samantha's hair. He visited her favorite salon and had her personal stylist show him how to do it. According to reports, his efforts did not produce the desired result!)

Fyodor did all of our haircuts, and at the end of mine, he always rapped me gently on the back of my head. He said it was a Russian tradition. I just think he was messing with me!

Question: How do astronauts get deliveries on the ISS? How often are astronauts able to request something be sent up for them, such as a certain brand of coffee?

Answer: Astronauts and cosmonauts on the ISS have multiple ways to receive cargo from Earth these days. The frequency may vary slightly over time, but roughly speaking, they tend to receive and welcome new cargo ships on a near-quarterly basis.

First and foremost is the old workhorse known as the Russian Progress. These solid ships have been delivering cargo to space safely and successfully for many years. Attractive because of their reliable perfor-

22. ISS Expedition 15 commander Fyodor Yurchikin (Russian cosmonaut) gives yours truly a haircut in the U.S. Node 1 module, Unity.

23. The Russian cargo ship, designated Progress 26, approaches an Earth-facing ISS docking port on the Russian segment.

mance and their ability to dock to the ISS automatically—once their delivered cargo is off-loaded—they are also loaded with unwanted trash and equipment that are no longer needed on board. Once stuffed to the gills, the ship will eventually undock and burn up in the Earth's atmosphere. It's sort of a spaceflight incinerator.

Since the retirement of the space shuttles—which were truly the "big dogs" of cargo delivery with their sixty-foot-long and fifteen-foot-diameter payload bays—other spacecraft have entered the game. It's an analogy similar to the trendy phrase used in modern-day football, "next man up!" Popular among the media and at NASA is SpaceX, Elon Musk's commercial spaceflight company. Using the company's *Dragon* capsule, NASA has hired SpaceX to deliver *and return* cargo for the ISS. The company's ability to deliver science experiments, data, and biological collectibles, such as urine and blood samples, back to Earth is key. So far, SpaceX is doing outstanding work.

Orbital ATK is another commercial company NASA has hired for delivering cargo to the ISS, but the company doesn't return anything to Earth. After a couple of successful missions, Orbital experienced a critical failure during its 2014 launch attempt, resulting in the loss of the Antares rocket and its cargo-carrying Cygnus capsule. Having since recovered from this incident, the company continues its successful efforts in the ISS cargo delivery business.

Finally, in the international consortium known as the ISS program, the European Space Agency and the Japanese Aerospace Exploration Agency (JAXA) both have demonstrated their cargo delivery capability as well. The former's Automated Transfer Vehicle (ATV), which I liken to a Progress on steroids, is nearing the end of its agreement with the ISS program for deliveries. Built with the help of the Russian Space Agency, the ATV is larger and brighter (i.e., more modern) than a Progress and just as dependable. The ATV was signed up for four deliveries, as I recall. JAXA, with its H-II Transfer Vehicle (HTV), had similar agreements for a limited number of cargo delivery flights. Unique to the HTV is its ability to use the robotic arm of the Japanese Experiment Module—also a JAXA contribution to the ISS—to help empty portions of its cargo load and place it on the module's external porch. Both the ATV and the HTV have successfully completed multiple cargo deliveries to the ISS.

It remains to be seen how the ISS program will one day try to get larger pieces of equipment—new external solar array batteries or thermal system pump modules, for instance—to the ISS. Prior to the shuttle program's demise, many spare parts were delivered to space and are now stored externally on platforms called integrated stowage platforms. "Living outdoors" in the harsh environment of outer space could pose potential problems when the equipment is needed to operate one day. We shall see.

Question: What does FE stand for on the ISS? On the daily time lines of the ISS, the crew members are labeled as FE-(number). What does this mean?

Answer: The abbreviation "FE" stands for flight engineer. The term had been used in the shuttle days, and the so-called mission specialist (MS)-2 was designated as the flight engineer on the middeck. In that position, the MS-2 helped coordinate the duties of the commander and pilot by serving as the procedural watchdog during the ascent, orbit, and entry phases of flight. The MS-2 sat in the seat directly behind the center console, putting him or her between and slightly behind the commander and the pilot. To my knowledge, the MS-1, who sat directly behind the pilot, did not have an FE designation.

In the ISS, the flight engineer is more prominent. Each Soyuz mission has a commander, a flight engineer-1, and a flight engineer-2. The FE-1 is a left-seater in the Soyuz and serves as a backup and "right-hand" man or woman to the Soyuz commander. I was a right-seater, or FE2, and had minimal duties with regard to my Soyuz crew. (I turned the suit fans on and off, and tried not to break anything.)

On the station crew's daily planning schedule, designations are for the current commander of the ISS, with all other crew persons having an FE designation. This enables planners to assign and monitor tasks being worked by individual crew members on the ISS on any given day.

Question: How is the lifting of the ISS carried out? During this process, do astronauts need to get in their space suits?

Answer: The lifting—actually, we call it boosting—of the ISS to a higher orbit is a periodic operation. Due to the drag experienced by the ISS because of its huge solar array area—the arrays act as big sails in the "wind," which we know as Earth's atmosphere—the orbit will decay, or get lower, with time. This is typically an undesirable characteristic.

The orbit of the ISS is essentially circular. (Yeah, I know most orbits are really elliptical, with very small eccentricities, but for the sake of all the non-rocket scientists out there, let's agree for this discussion that they are circular, shall we?) As the station moves in that circular orbit day after day, month after month, it is constantly "falling" from the sky, albeit slowly. As the solar arrays rotate toward the sun and generate much needed electricity, the area presented to the drag-laden atmosphere varies. (Think of extending your hand from the window of a moving car. If you put your palm into the air flow, your hand will shoot back, pushed by the air. If you turned the edge of your hand into the flow, you can maintain its position more easily.) The resultant overall effect experienced is that the ISS moves more slowly. As it slows in orbital velocity, the force of gravity exerts more influence over the huge ISS mass, pulling it lower and lower toward Earth. We must counteract that occurrence by raising the ISS back up into a higher orbit.

How do we raise the ISS orbit? We use Russian vehicles such as the Soyuz capsule or the Progress cargo vehicle, and when they were formerly available, we used the European Space Agency's ATV and the U.S. space shuttles. (Now the shuttle program's vehicles are on blocks in museums around the United States.) Most often positioned appropriately on a docking port at the back end of the ISS, we briefly fire the Russian-made orbital maneuvering engines to give us just enough added velocity to raise the ISS orbit. This added thrust must be in the station's direction of orbital travel, creating what rocket scientists call a posigrade burn and positively raising the orbit.

While this sounds pretty dynamic and exciting, for the astronauts and cosmonauts living inside the ISS, it was a bit mundane (at least for me anyway). If we weren't too busy and knew when the burn was coming—the control centers usually forewarned us—we could do stupid astronaut tricks, such as placing oranges or Nerf footballs at the front of a station module and watch them fly to the back as the burn accelerated the ISS in the forward direction. No space suit donning was necessary as the burns were typically of low thrust and short duration. In other words, it didn't take much.

These techniques can also be used to move the ISS away from potential debris that may be encroaching, or flying a bit too closely for our comfort, on the ISS structure. Posigrade (raise the orbit), retrograde (lower the orbit), or radial burns (shift the orbit) of various magnitudes and combinations can be executed, depending on the need and the location of the offending debris, but that's another class session.

Question: When astronauts are in space, can they still be blinded by the sun? Do they have to avoid looking at it?

Answer: Absolutely we avoid looking into the sun just as we do on Earth. As a matter of fact, we often wear sunglasses to shield our eyes during docking operations when an untimely appearance of the sun in our eyes could be detrimental. During space walks, the suit we wear has a golden helmet visor specifically designed to protect our eyes from the brightness of the sun while we are working outside the International Space Station. Remember too that we are not looking through the Earth's atmosphere. We are flying above it, and that gives us a tad more brightness than you might experience on the planet. Most of the ISS windows are provided with UV shielding, further protecting us from the more harmful aspects of the sun's rays.

Question: For astronauts, what is your most memorable spill of a harmless substance on the International Space Station (and it doesn't have to be that *you* spilled it)? What kinds of harmless and annoying things have gotten loose in microgravity?

Answer: My answer is simple:

1. Butterscotch pudding, which ended up on my station commander Fyodor Yurchikin, and
2. Hot Kona coffee, which doused the door to my crewmate Oleg Kotov's sleep compartment.

My spills were due to inattentiveness. With the pudding, I didn't realize I was putting pressure on the bag when I cut it open, resulting in a squirt of pudding on Fyodor's shirt and arm. With the coffee, I had swooped down to help Oleg near the deck and turned away from the device that was filling my coffee bag. I failed to see it was overfilling and rapidly at that. Ultimately the bag sprung a leak, spewing a steady stream of Kona coffee with cream and sugar all over his door. Cleanup was relatively simple in those two cases, but it can be a lot messier in the zero-gravity environment, trust me.

Harmful substances are rarely spilled as astronauts typically don't interact purposely with anything onboard that is dangerous. If they do, tremendous care and proper equipment combine to help keep everything—and everyone—safe.

Question: Are there any instances when two space shuttles were deployed at the same time?

Answer: Two space shuttles deployed at the same time? Nope. Only in the movie *Armageddon*!

NASA and the space shuttle program did have a launch on need (LON) effort, designed to launch a space shuttle into orbit, in the event something had gone wrong elsewhere in space. This effort was developed only to a certain level of detail: A crew was named, tentatively, and personnel received rudimentary training to be ready to go. These crews had flown previous missions, so the training was essentially a refresher of sorts. But none of this was ever put into action.

Many felt this capability should have been used with the *Columbia* accident. But during the mission (STS-107)—according to some—available resources such as satellite imagery were not utilized that might have shown (key word is "might") there was an issue with the shuttle's reinforced carbon-carbon tiles. In addition, it was often speculated that nothing could have been done to save the crew, even had NASA known of the true situation.

Postflight analysis of the tragedy showed foam released from the external tank during launch had struck an area of the tiles on Columbia's left wing, creating a not insignificant hole. That opening in the wing's leading edge eventually allowed extremely hot atmospheric gases to reach the internal aluminum structure, compromising the shuttle wing's integrity and leading to its destruction on reentry. But in essence, NASA didn't really know there was an issue until it was too late.

To have utilized the LON capability, we would have needed to know there was an issue while the shuttle was still in a stable orbit. If LON had been utilized, it remains pure speculation as to whether a rescue attempt would have been successful, as no practiced and verified methodologies existed at the time for an effort of this magnitude.

Much has been documented on this terrible tragedy, and I do not pretend to have absolute answers, despite the fact I was an active astronaut and crew family escort at the time. I urge all to consult available resources for more definitive and detailed information, including chapter 7 in *The Ordinary Spaceman* that is dedicated to this mission.

Each year in the January–February time frame, when all of NASA reflects on the anniversary of this tragedy—and the *Challenger* and *Apollo 1* disasters as well—I can still respectfully ask that we all keep lookin' up. But this time, please say a silent prayer of thanks for those brave souls who gave their lives for our space program. They were my heroes, my colleagues, and my friends.

Question: Can astronauts return to Earth from the ISS without Russia? Suppose President Donald Trump does something incredibly dumb and pisses off Vladimir Putin, resulting in Cold War II, so the two Soyuz spacecraft attached to the ISS are now off-limits. Are the astronauts stuck on the ISS forever? Can Elon Musk save the day?

Answer: Can American astronauts return to Earth from the ISS without Russia? Nope. No way, no how—at least not today.

The United States mothballed the space shuttles, all of them. As noted previously, they are in various museums across the country. *Discovery*, *Atlantis*, *Endeavour* all of them are decommissioned. Bad move, I'd say. Between the three of them, they had over 150 flights left in 'em.

So now here we are with the shuttle retired, and we must rely—once again, as we did after the *Columbia* tragedy—on the Russians and their solid, reliable Soyuz spaceship-rocket combination to ferry our astronauts to and from the ISS. And make no mistake about it; we pay them handsomely for the privilege. Further, this will be the case for the foreseeable future until Boeing or SpaceX, or both, can ready their systems for safe, human spaceflight. Here's hoping they do just that and soon!

Question: What are the greatest dangers to the ISS?

Answer: The greatest dangers to the ISS are few. There are three big ones to be precise. When living in an aluminum can, now some 250 nautical miles above Earth, it is very bad form to ever have any of the following:

1. A fire,
2. A depressurization, or
3. A toxic spill.

To qualify each of those a bit, a fire is pretty easy to understand. Flames are bad. We have extinguishers to put them out, hopefully, with the U.S. side using CO_2 and the Russians using soapy water. Neither is allowed to be used in the other's segment. The U.S. version has nozzles for "squirting" behind panels—the most likely place where a fire due to faulty wiring will occur—using what are called fire ports.

A depressurization means you are losing air. It could be from a meteor strike, a leaking seal somewhere, or even a hit from another spacecraft as happened to the Russian Mir station many moons ago. All are serious and may require reasonably swift action. For this type of event, the bigger the leak (read as "hole"), the faster the air loss and the direr the situation. Astronauts and cosmonauts, working in tandem with the ground control teams in their respective Mission Control Centers, must first gather together in a safe haven—usually the Russian service module—and figure out the reserve time (T_{res}), or the time until the crew must abandon the station. Once T_{res} is known (and the ground team agrees) and if the T_{res} is large enough, then the crew members perform a methodical search for the leak. Their first step is to isolate the Russian and U.S. segments, always keeping their Soyuz escape vehicles accessible by not closing a hatch behind them that takes away access. By dividing the station in half, they can quickly deduce which segment—Russian or U.S.—is leaking. Once that is known, they continue methodically closing but not locking hatches until the leaking module is isolated. As they move the hatches toward a closed position, they note whether the hatch is being sucked closed or resists closing. This gives them a clue as to which side of the hatch the leak is located. Again, if enough T_{res} remains, then they can remove equipment that they want to keep from experiencing depressurized or vacuum conditions.

If the leak is large and the T_{res} is small, then the crew members head immediately to their respective Soyuz capsules and prepare to undock and return to Earth, abandoning the ISS.

If the leak is found and isolated, and things have calmed down a bit, then they can later work with ground control to develop plans to repressurize the module, to enter it, and to attempt to patch the leak with what really amounts to a Silly Putty–type patch kit. Not very extravagant, it's effective nonetheless.

A toxic spill is also quite serious, as it typically deals with ammonia. Nasty stuff used to cool the outside part of the station's electronics, it flows through heat exchangers that interface with the water used to cool the ISS interior (U.S. side). (The fluid tubes in the exchanger are adjacent and heat is transferred from the warm flow to the cold flow.) Since the two fluids (water and ammonia) interface in the hull of some ISS U.S. modules, the potential for a leak into the station's interior is possible and dangerous. If the quantity of ammonia inside is high, then this nasty situation calls for us to execute what NASA calls **bold face**, or memorized, procedures.

When I lived there in 2007, **bold face** dictated that if we smelled ammonia or the toxic spill alarm went off, then we were to immediately don oxygen masks and push the alarm button to initiate the ISS automatic shutdown procedures or at least ensure they had already been initiated. After that our task was to immediately evacuate the U. S. portion of the station via U.S. Node 1 and move all the way to the Russian FGB module. We would be, in essence, trying to outrace the ammonia cloud. Once inside the Russian FGB module, and after closing the Node 1 aft hatch ASAP and isolating the ammonia to the U.S. side, we were also instructed to doff all clothing as it could be contaminated with ammonia, especially if we had to fly through an ammonia cloud. This would make for some very embarrassing moments, depending on crew composition, but I'm guessing in this emergency no one would care!

Many other things could go wrong—power failures, loss of control authority, or lost communications with the ground—but the odds are low. And their seriousness pales in comparison to the big three outlined here.

Question: How does NASA manage odors inside the International Space Station?

Answer: The following information comes from Robert Frost, who is a NASA engineer.

Odors can result from equipment off-gassing, crew metabolic processes, food, experiments, and returning EVA [extravehicular activity] crew members. 216 such contaminants have been identified and designed for.

In the SM (Service Module), the Micro-purification Unit (BMP) provides a regenerable means to remove both low and high molecular weight contaminants.

In the Lab, the Trace Contaminant Control Subassembly (TCCS) performs a similar function.

Both of these units are nominally operating. Either one is capable of providing the trace contaminant removal for the entire ISS. The BMP vents contaminants overboard while the TCCS traps them in replaceable beds.

Major components of the TCCS include an activated charcoal bed, a catalytic oxidizer assembly, a lithium hydroxide [LiOH] sorbent bed, a fan, and a flow meter. Although the TCCS removes most atmospheric contaminants with the charcoal bed, the high temperature catalytic oxidizer is required for removal of lower molecular weight compounds, such as methane.

TCCS inlet air is drawn directly from the open cabin atmosphere into the activated charcoal bed by the TCCS fan, which is downstream of the charcoal bed. The charcoal bed is impregnated with phosphoric acid, which enables it to absorb ammonia. Downstream of the fan, the process air is split into two flow streams; one going to the Catalytic Oxidizer Assembly, and one to a bypass line. The flow rate to the Catalytic Oxidizer Assembly, measured by a flow meter, is used to control the speed of the fan to provide a specified rate of flow to the oxidizer. The remaining flow is sent through the bypass.

A regenerative heat exchanger and a resistance heater are used to heat the air entering the oxidizer to approximately 400° C. The catalyst oxidizes the organic compounds to CO_2 and water, and converts the inorganic compounds to acidic gases such as hydrogen chloride, hydrogen fluoride, and sulfur dioxide. Air leaving the catalyst bed is cooled in the regenerative heat exchanger and is circulated through the LiOH sorbent bed. This bed removes any acid by-products produced during the oxidation process. The air exiting the sorbent bed combines with the unprocessed bypass air and returns directly to the ducts.

One of my favorite things to do is to piggyback on the wonderful—and always technically sound—answers from NASA's Robert Frost. While he gives you the clear and concise mechanical, textbook-type answer to the question, I can jump in with some real-life experiences. Together, it is my hope that our "tag team" effort gives you some insight that you may not have expected but can honestly appreciate and enjoy. Hopefully, we are all learning something together.

As a two-time space flier with over 167 days in space—with all but 8 on the ISS—I have definite personal knowledge of the odors existing inside the station's anodized aluminum hull.

The systems Robert mentioned typically perform flawlessly, with the crew doing routine maintenance on those systems per the schedule provided by the ground. This work, in and of itself, gives us a significantly odor-free environment on the ISS. That's not to say we are without smells, however. Our standard atmosphere, composed of the very same percentages of constituent gases that we have here on Earth, is regulated to 14.7 pounds per square inch and gives us a shirt-sleeved environment maintaining about 72 degrees Fahrenheit, 55 percent humidity, and with a slight breeze out of the south!

Oleg Kotov, my Expedition 15 Russian crewmate and Soyuz commander, liked to stash his used workout clothes above the forward-facing FGB hatch. This was not my favorite choice for stowing sweaty workout gear as there was not a very good chance that it would dry out effectively. I chose to put my nasty shorts, socks, and T-shirt onto a handrail in the U.S. segment's Node 1 module. This handrail was near an air-conditioning vent, meaning fresh, cold air would blow across my sweaty laundry for many hours until I donned them—dry as a bone—the next day. Decreasing their ability to generate any locker-room odors, that special placement also allowed for our environmental systems to easily soak up my sweat and turn it into drinking water for later.

Food odors were also present, but they didn't seem unusually overpowering to me.

Eating a fish dish often produced the most pungent odor, especially the U.S. version of seafood gumbo. It might take a couple of hours to purge that smell from the airflow of the ISS. On shuttle missions, many commanders outlawed the eating of seafood gumbo due to its distinctive, and disliked, smell.

And don't worry too much about the stink from the toilets. The airflow systems there, especially in the U.S. segment but not so much for the Russian side, were very effective. They pulled the stench quickly and completely into the bowels (from one set to another) of the ISS, where they were absorbed efficiently by filters. (The shuttles had the absolute best system for containing poop odors.)

Finally, there is the "smell of space." Oft mentioned by astronauts in answering insightful questions from folks like you, the smell of space is somewhat hard to describe. Ever distinct—I would know it instantly if I smelled it again—it has been likened to smells associated with welding or the burning of ozone. (Now who the heck *really* knows what that smells like?!) It was most noticeable following a space walk, when crews and their equipment returned to the inside of the ISS. I remember being able to smell traces of this unique scent for several days following an excursion into the unforgiving vacuum of space.

Question: What is the speed of the internet/Wi-Fi that astronauts can use to connect with their personal gadgets on the ISS?

Answer: Slow . . . and slower! When I returned to the ISS with the STS-131 *Discovery* crew in April 2010, social media had just begun to take off (good space reference, huh?). When I arrived, I hoped to do some tweeting from space. I tried to use the internet capability from JAXA astronaut Soichi Noguchi's sleep station, but it took so long I gave up.

I resorted to sending tweets via email to the ground—a much faster process although mail buffers only synced about twice per day—and then the email was converted into a tweet that was sent throughout the Twitterverse (@Astro_Clay). Not perfect but functional. However, I could not read or review tweets in real time unless they were converted into an email.

Today on the ISS the internet/Wi-Fi capability is just a bit faster. But given that the process has many steps, it's no wonder it takes a while. My NASA contacts tell me that the real bottleneck in all of this is the interface between the station's laptop computer and the computers on the ground. There's a delay of between 500 and 700 milliseconds in that connection as compared to the 12 to 25 milliseconds we might experience here on Earth. That dictates an overall speed of about 128 kilobytes per second, which is only slightly faster than a dial-up one (56 kilobytes per second). So compared to what we are used to with today's internet, the internet/Wi-Fi capability on the ISS is still as slow as molasses in January.

Question: Why do astronauts use a pencil instead of a pen in space? What pens do astronauts use in space today?

Answer: I don't think it's correct to assume that astronauts use pencils instead of pens in space. You see, writing utensils in space are many and varied. They range from mechanical pencils to ballpoint pens to highlighters and Sharpie markers in an assortment of wonderful colors. (Different colors aid astronauts in the psychological department, brightening a sometimes mundane place.)

During my time on board the ISS, I was never without my trusty Sharpie. We had the new ones that worked just like a clickable/retractable ballpoint pen. I clipped it to the collar of my T-shirt or, when wearing nicer clothes, put it in a blue knee bag that was Velcroed to the leg of my pants.

Being from Nebraska, my Sharpie color of choice was—you guessed it—*red*. So when you watch video from the ISS and you spot anything that appears to have been labeled with a red Sharpie, it probably was my handiwork. I'm almost certain of that. (I labeled lots of the white cargo bags the astronauts haul around, so check those out during their space walks.)

Fisher Space Pens are also used quite often on orbit. I Velcroed them to the workstation that sat outside my sleep station in the U.S. Destiny Laboratory. They work quite well in space due to their ink-pumping action. It's my understanding that they are famous for their reliable use in outer space.

In fact, there's an old joke (myth) about how the United States spent thousands of dollars to develop the Fisher Space Pen specifically so astronauts could use them to write in zero gravity. But as a matter of fact, the U.S. government spent nothing; Fisher developed the pen using its own finances. Its fancy design allows ink to be pumped to the writing element and surface, enabling a person to write in any orientation. As for the Russians? Well, now, they just used a pencil.

Mechanical pencils with thicker leads than number 2 pencils were also provided. I used them as well but could get frustrated by the lead's tendency to break frequently. I was not worried about tiny pieces of lead getting into my eyes. You'd have to work pretty hard to make that happen, and the ISS ventilation system was quite good at sucking debris of that magnitude out of the air and out of harm's way.

Many mission controllers would argue that, due to my mental state, I should have been using Crayola crayons. Go figure.

Question: Do 3D printers work in microgravity and zero gravity?

Answer: Apparently, NASA believes they do. A 3D printer is already installed and working on the ISS. Numerous individual items have already been printed including a ratchet wrench, without the ratcheting function, and a multipurpose tool that fits in an astronaut's pants pocket.

As a layman in the 3D printing regime, I can only speculate as to its future uses and possibilities for success, but in terms of re-creating spare parts or tools, it definitely has merit.

Perhaps they can "print" a 3D coffee cup to hold their espresso when it comes out of their spiffy new beverage machine piping hot!?

Question: Is it possible to trigger the ISS fire/master alarm manually?

Answer: Yes, it is very possible! We can trigger the caution and warning system manually for all emergencies: for a fire, for a depressurization, or for a toxic spill.

The process is simple. Smell smoke, see flames? Push the *fire* button. Feel your ears equalize due to a change in pressure? Check with your crewmates to see if they felt it too. If they did, it's a pretty good chance there's a leak somewhere, so push the DP/DT (PRESS-depressurization) button. Smell ammonia or see some other nasty liquid/gas floating in the ISS, push the ATM (toxic atmosphere) button!

If it is a real emergency, many other steps are taken based on what's happening. As noted earlier, if it's an ammonia leak, then you will also don an oxygen mask, head toward the Russian segment, close the U.S. aft hatch on the Node 1 module, and get naked. Perhaps that seems a bit extreme for coed crews, but ammonia is nasty stuff and can be in your clothes if you flew through the cloud. Ingesting even a small amount can be fatal.

A fire may also lead to your donning an oxygen mask and discharging an extinguisher. A severe depressurization may have your crew bolting for the Soyuz "lifeboats" in preparation for an early, but expedited, departure for home.

Astronauts and cosmonauts practice these three serious emergency scenarios many times. We practice individually and then many times with our crewmates. All must know the procedures, roles, and responsibilities—in addition to their limited ability to rely on the MCC—to maximize the possibility of a successful outcome.

Alarms may be activated manually through a laptop computer display or by flying to a U.S. caution and warning panel, lifting the protective clear plastic cover, and pushing the appropriate button: fire, DP/DT, ATM. Alarms may also be manually activated from various panels in the Russian segment and their laptops.

A serious emergency on board the ISS has been, thankfully, a rare occurrence. There have been some false alarms along the way, but given that the ISS has been sailing around our planet with humans in tow since about 2000 or so, the safety record has been outstanding. Sticking with the adage that "proper planning prevents poor performance," the ISS crews and ground team are ready for anything. And that's a very good thing.

9

Space Suits and Space Walks

Question: How many astronaut suits are in use now?

Answer: Interesting question, with what I think is a pretty simple answer. But we have to narrow the field a bit. Let's talk American astronauts' suits first. Depending on your definition of "astronaut suits," we can come up with several answers. If we eliminate those suits purchased at fine clothier establishments around the world, we can focus on what I believe is the true nature of your question.

The white extravehicular mobility unit (EMU) is one of the most well-known American space suits. Designed for astronauts to wear during space walks, the EMU is essentially a spaceship for the astronaut. Containing everything an astronaut needs to survive in the vacuum of outer space, the EMU is a marvel of engineering design and functionality.

The suit that most recognize is the advanced crew escape system (ACES) space suit. Designated by many as the orange "Pumpkin Suit," this protective garment was worn during the ascent and entry phases of space shuttle flights—that is, going into space and coming back down to Earth—and came into being after the *Challenger* accident in 1986.

While not as extravagant as the EMU, the ACES suit was designed to give astronauts a safe haven capability for a short time in the event they needed to bail out from the space shuttle (recall the *Challenger* and *Columbia* disasters). The ACES suit didn't exhibit all characteristics of an EMU—for example, it didn't have multiple layers of material to protect astronauts from meteorite impacts—but it is currently being evaluated for potential use as a spacewalking suit with the Space Launch System and Orion programs.

One final space suit that American astronauts use is the famous blue flight suit. Made of fire-retardant Nomex fabric, it is typically worn when flying in the T-38 Talon—an air force training jet adapted for use by astronauts—and when astronauts make public appearances. Easily lending itself to the astronaut nickname "blue suiters," it is one of the most recognizable uniforms worn by American, and oftentimes Russian, space fliers.

It's a different game in Russia. First off, all of their suits are named for birds. The primary suit of the Russians is the Sokol suit, which is analogous to our ACES suit and whose name in English means falcon. Worn during all phases of flight in the Russian Soyuz spacecraft, the Sokol is a single-piece suit. Everything is connected, except for the gloves.

The Russians also have a spacewalking suit called the Orlan, a one-size-fits-all type of suit that you can adjust in the length of the arms and legs a bit, and with only two sizes of gloves. The Orlan (English translation is eagle) allows you to don it all by yourself by climbing in the back door and closing said door all by yourself! Self-donning a spacewalking suit is a much different situation than with an American EMU, which requires a minimum of two crew members to perform the task of donning. Another interesting Russian suit is known as the Penguin. Designed to help astronauts and cosmonauts maintain their musculature, the Penguin suit has internal cabling and adjustable belts that can be tightened or loosened; they're intended to tense a space flier's muscles and bones in microgravity to help keep him or her fit. Think of it as a suit of isometric toning. Many American astronauts (me included) cut off all the straps and wore the Penguin suit for a few publicity shots, never donning it again.

Finally the Russians have their Chibis (Lapwing; a shore-type wading bird). Not necessarily a suit, it's more like a fancy pair of trousers. Often the brunt of various *Wallace and Gromit* jokes, the Chibis is also known as a lower-body negative pressure suit. It is donned by Soyuz fliers during the period a few weeks prior to returning to Earth. Tightly sealed around the cosmonaut's lower body, it uses pressure to stimulate a cosmonaut's leg muscles in hopes of preparing the fliers for the return to—and first steps on—Earth.

And, oh, there's one other suit that I donned with regularity while living aboard the ISS usually after exercise. It's called a birthday suit, and, no, I don't have any pictures of that one.

Question: Why do space suits require an airlock to be stored?

Answer: Once again my NASA colleague and friend Robert Frost hits another home run. His answer, brief and succinct, is also totally correct when he says, "They don't."

Piggybacking on Robert's genius once again, I believe I can offer some additional insight into how our space suits are stored.

First, understand that we are talking about our spacewalking, or extravehicular activity (EVA), suit. Also known as the EMU, you may have seen pictures of these white EVA suits being stored in the ISS airlock.

Hanging on the airlock walls, the suits face each other, with their appendages magically suspended in the microgravity environment. These suits are prepped for members of the crew, specifically sized for the designated spacewalkers, and ready to go out the door when needed.

Sometimes we also have other suits on board. Perhaps containing new technology, parts, and software upgrades, they are being readied to replace older, and maybe malfunctioning, suits soon headed for home. We store these not-yet-ready-for-prime-time suits in a bag of sorts. Cleverly designed and adequately engineered, it is composed of canvas, straps, buckles, and pockets that work in tandem to help the onboard crew bundle up the extra suits to minimize their footprint in our living and working space.

Think of one of these extra space suits as being outfitted in a standard kitchen apron. With the suit's helmet, gloves, and boots attached, the "apron" is draped over the front of the suit and covers the arms, which are folded over the chest. The legs and boots are then smooshed up beneath the upper half of the suit—the hard upper torso—while the apron is pulled under them toward the back. As straps are tightened and the communication cap is placed in the apron's front pocket, the suit is a smaller—and more manageable—package ready for stowing somewhere outside the already occupied airlock.

To get an idea of the resultant bundle's size, imagine an NFL football player clad in all his gear curling up in a ball with his arms holding his knees tightly against his chest. While still of formidable size, he takes up much less space than our dangling "suits-at-the-ready."

Question: What does it physically feel like to be inside a space suit?

Answer: To answer this question specifically, I am assuming the person asking is *not* talking of the "mental joys" of wearing an astronaut space suit for the first time, but, yes, in that vein, it can be euphoric. I'm guessing the questioner wants to know what it really feels like physically.

Well, wearing an astronaut suit is damned uncomfortable. And that comment applies to any and all of the astronaut suits we need to wear, at least in this astronaut's opinion.

Most are familiar with the orange Pumpkin Suit, officially known as the ACES suit. This monstrosity—composed of the boots, the gloves, the helmet, and the suit, its pockets filled with survival gear—was originally designed to keep us alive in the event of a survivable bailout from an unsafe or unflyable shuttle. Today it is being modified and tested for use with NASA's Orion Program, including an evaluation of its potential as a spacewalking suit. With an emergency supply of breathing oxygen of about ten minutes' worth, the original ACES suit could be pressurized to aid an individual bailout, followed by a parachute descent into the ocean or land, but keep in mind the surface of the Earth is about 70 percent water. Colored bright orange for enhanced visibility, the suit's contrasting white helmet is covered in reflective tape in hopes of making tiny individual floating astronauts look as large as possible to the search and recovery teams scanning vast ocean surfaces.

Heavy, bulky, and quite warm, the suit itself is not the only thing contributing to the astronaut's discomfort. Beneath the suit we wore a maximum absorbency garment, or diaper; long underwear (Patagonia brand, typically); thick socks; and a liquid-cooling garment. But comfort is not the concern; astronaut survival is. We needed all that gear to help keep us alive at altitudes high in the Earth's atmosphere—ten thousand feet is the design-preferred bailout region—or in the frigid and turbulent waters of the North Atlantic Ocean.

As noted, the ACES could soon be making a comeback, serving in a modified EVA-configuration adapted for use with the Space Launch System and Orion programs. Stay tuned.

More familiar to most may be the bright white spacewalking suit, the extravehicular mobility unit. While "extravehicular" may be descriptive of its purpose for going outside the ship, "mobility" is a bit of a stretch.

Worn during excursions outside the shuttle or ISS, this suit can weigh upward of 350 pounds on Earth. Assembled inside our space vehicle, the EMU comprises boots; the lower torso assembly, or legs and "briefs"; the hard upper torso, or "shirt" and arms; gloves; and a helmet. Attached to the back of this extremely rigid suit is a jet pack known as the simplified aid for EVA rescue (SAFER). If one were ever unfortunate (or stupid) enough to fall off the ISS, this jet pack would be the astronaut's only hope of survival, giving the astronaut a one-time shot to fly "Buck Rogers style" back and grab a handrail on the station's exterior.

Much like its ACES cousin, the EMU requires the astronaut to don a diaper, a pair of long underwear, and a similar liquid-cooling garment. Comfort gloves—worn to lessen the chafing and pressure points of manipulating working gloves thousands of times during a space walk at 4.3 pounds per square inch absolute of pressure—seem "dainty" when compared to the rest of this multilayered cocoon. The EMU's white color helps to reflect the heat from the sun's rays, and a helmet adorned with headlights and a sun visor (with some actual gold in it) allow the astronaut to function during light and dark passes as we sail around the Earth at 17,500 miles per hour. Red-hashed stripes can be affixed to the legs and shoulders of the suit, giving the ISS internal crew and Mission Control team members the ability to tell who's who during scheduled six-plus-hour strolls "outside."

Functional, dependable, and incredibly well designed by NASA engineers, the suits did their jobs for me six times in space, and I wouldn't change a minute of it. But they were still damned uncomfortable.

Question: How does a space suit work?

Answer: How a space suit works depends on the kind of space suit. Describing exactly how they work would take way more space than I have here. Let's look at the question in a more general sense by seeing what, exactly, we need the suit to do.

Most space suits have very specific purposes. First and foremost, they must keep the wearer alive. They serve—in actuality—as individual custom spaceships of a sort and are designed for considerably differing environments.

In today's space program, as discussed previously, the United States uses two primary space suits. One, the advanced crew escape system suit is the familiar bright orange contraption that television viewers saw crews wear during their waving walk out to the Astrovan that took them to the shuttle launch pad. No longer needed in their current form with the end of the space shuttle program (boo hiss!), they are now being evaluated for use with future programs and being modified to perform space walks in the event of a capsule emergency—for example, should a pesky solar panel not deploy.

Designed to protect the crew only during the ascent and entry phases of flight, the ACES suit primary purpose was to provide a brief oxygen source, temperature control, and pressurization for a relatively short time in hopes of enabling the crew to transition through the upper atmosphere and parachute to safety in the event of a bailout emergency. As noted in chapter 8, this very purpose went unfulfilled during the *Columbia* (2003) tragedy and was not even available during *Challenger* (1986).

The spacewalking suit, also known as the extravehicular activity suit and being easily recognizable to even the most casual space buff, is bright white with identifying stripes of solid, dashed, and angled, or hashed, red and white. This suit truly is an astronaut's spaceship. Worn outside of a vehicle while working in the vacuum of outer space, this suit's purposeful design is a bit more extravagant than that of the ACES suit.

The EVA suit—with its hard upper torso, gloves, helmet, lower torso assembly, and attached boots—can be an astronaut's "home" for periods upward of eight hours or more. While astronaut survivability is its foremost requirement, it must also allow the crew member to perform meaningful work. As such, its design must be considerably more robust. While meeting the standard need for an oxygen supply—much longer than ten minutes!—the lengthy time spent in the suit also requires a means of removing CO_2, body heat, and humidity. Throw in temperature control for times when the sun is (and isn't) visible and the resulting swing ranging from about 250 degrees Fahrenheit to minus 250 degrees

Fahrenheit, this suit needs complexities such as pumps and fans for air and water circulation, filters, and batteries.

Further, when working in the darkest portions of orbits due to the Earth's blocking the sun, headlamps and television cameras add to the EVA suit's capability. Dodging micrometeorites isn't really an option during a space walk, so layer upon layer of fancy materials such as bladders, insulation, and Kevlar help ensure our American heroes will indeed make it back inside their comfy spaceship. Top it all off with a bag full of water for the tired and very thirsty astronaut to drink and a tool-storage belt to carry his or her wares, and this suit is the Cadillac of all space suits.

While not totally answering your question, it's obvious that the workings of any space suit are complex, and the astronaut's life depends on it working at 100 percent all the time.

Question: How dangerous are extravehicular activities? I'm attending the Space Apps hackathon in Paris, and I'm planning on designing the prototype of a drone with an electrically powered spacecraft propulsion system to assist or replace astronauts during EVAs. I'd like to know if EVAs are dangerous enough to make my project appealing.

Answer: Your project has tremendous appeal. At least it does to me, the Ordinary Spaceman!

Having ventured outside the space shuttle and the ISS a combined six times during my fifteen-year astronaut career, I found EVAs—otherwise known as space walks—wonderfully challenging and rewarding. With that being said, they can be extremely dangerous endeavors.

NASA and its very competent army of scientists and engineers will do everything in their power to reduce the danger for any astronaut who ventures outside into the unforgiving vacuum of outer space. While mishaps during space walks have been few and far between, there is definitely a role to be played by robotic machines, including your drone concept. As a matter of fact, NASA continues to work on these types of ideas, many similar in concept to yours. The key is that spacewalking humans will not be replaced, but the risk of exposing them to danger can be reduced, in part, through the use of robotic machines performing many of the more routine and mundane spacewalking tasks.

Robonaut, a *huge* robot design project between General Motors and NASA, is hoped to be exactly that—a robotic assistant to perform EVA tasks, thereby not requiring that a human go outside to do them. However, Robonaut remains stuck inside the ISS, legs and all, awaiting the time when its designers can advance its development significantly to allow it to successfully perform work outside of the ISS with minimal human interaction. As a Mission Control Center capsule communicator back in 2013, Robonaut—already "living" on board the ISS—failed to pull a single tissue from its packaging. It was not a good omen for a robot hoping to positively contribute to maintenance and repair excursions outside the station. But we'll get there eventually, of that I am certain.

The Canadian Space Agency also has an entry into the world of robotic space walk helpers. Originally called the special-purpose dexterous manipulator, it was nicknamed Dextre when management tired of people pronouncing the acronym "SPDM" because it sounded as though they were discharging a huge ball of phlegm from their throats. Dextre attaches to the Canadarm2 robotic manipulator and has already demonstrated potential in the realm of removing and replacing standard orbital replacement units on the station's exterior. However, Dextre is slow and methodical, which is not necessarily a bad thing, but it's so slow and methodical that sending a human outside to do the space walk is probably faster at this juncture. In addition, Dextre still requires humans to manipulate its actions and the robotic arm in real time, reducing a bit the rationale for using robots in the first place—that is, increasing astronaut safety first, as well as efficiency and cost of operations.

Finally, a group of students from Massachusetts Institute of Technology—paired with NASA and industry partners—participate in an ongoing ISS effort called Synchronized Position Hold, Engage, Reorient Experimental Satellite (SPHERES). First on board the ISS in 2006, three bowling ball–sized satellites perform formation flying and other precision operations, including docking with one another. The SPHERES now have a camera attachment, which I hope will soon allow them to head outside the ISS and perform much-needed automated video surveillance of the ISS, helping us to find issues before they find and surprise us.

The future is coming. I just want it to get here much faster than it is.

Question: How do astronauts eat during a space walk? I read that Clayton C. Anderson's first space walk was seven hours and forty-one minutes long. Do astronauts go without food for the duration of the space walk?

Answer: How do astronauts eat during a space walk? We don't! The American spacewalking suit, the EMU, only has a place for a small drink bag, which is filled with water. To combat hunger, astronauts try to eat a solid breakfast before donning their suit. This can prevent some issues, as many of us are so excited to go outside that our appetites are a bit suppressed. For me, it was similar to when I played high school basketball. On game day, I was always a bit nervous as I wanted to play my absolute best. I would come home from school and eat two pieces of toast with butter. That was it. I'd be starving by the time the game was over! It was the same for me doing space walks.

I remember one of our ventures outside during our STS-131 mission on *Discovery*. Fellow astronaut Rick Mastracchio and I had completed our walk, which did cover about seven-plus hours (google it!). I hadn't eaten since I tried to choke down some peanut butter in a tortilla (yum!) and a PowerBar (yuk!) that morning. After I doffed my suit, cleaned up and got dressed, and returned to the airlock, my former Expedition 15 crewmate Oleg Kotov—now the ISS Expedition 23 commander—floated over to me with a freshly heated can of Russian lamb with vegetables and a spoon. He had remembered my favorite meal from our five-month mission in 2007. Now that's good "expeditionary behavior!"

Question: How do astronauts deal with an itchy body part while inside the protective suits in space?

Answer: If you have an itchy body part while in your protective space suit and working outside in the vacuum of space, what do you do? You "shake, rattle, and roll," baby! That's the only thing, really, that you can do to alleviate that itch. Hopefully, between the incredibly bulky and stiff suit itself and the liquid cooling garment you are wearing beneath the suit, you can wiggle your body enough to effectively scratch that itch.

However, if it's in your helmet and facial area, it's a bit more difficult. Then we usually resort to enlisting the help of what we call the Valsalva device, a piece of foam attached with special "space-allowed" adhesive to the lower front rim on the inside of your helmet. At one time the Valsalva device was cube shaped with a slot cut down its middle, allowing you to bury your nose in the slot and blow gently to equalize the pressure in your ears. This process helped as you sank into the increasing pressure of NASA's Neutral Buoyancy Lab or patiently waited in the space shuttle or the ISS airlock as the pressure decreased during the depressurization to start your space walk and then increased while repressurizing at the end of your EVA.

With the help of veteran astronaut and veterinarian Rick Linnehan, among others, the foam cube was modified into a shape reminiscent of the Valsalva device first used during Apollo. Shaped similar to the knuckles of your index and middle fingers when they fold at the lower joint, you moved your head down toward the foam sculpture, inserted the "knuckles" *inside* your nostrils, and pushed down gently, giving you a nice seal for equalizing the pressure. Both devices also served as won-

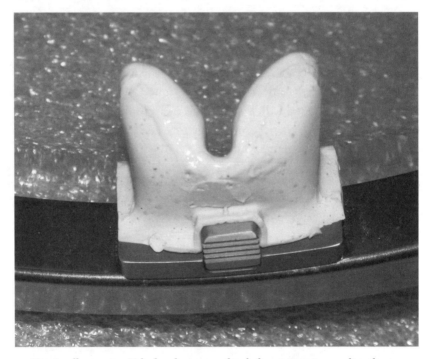

24. The Apollo-version Valsalva device, used to help astronauts equalize the pressure in their ears, is clearly visible attatched to the helmet base ring.

derful "scratch pads" for those nuisance itches one always seems to get when one cannot scratch them. In the event of a really bad sneeze, you could also use the device to wipe some of the discharged snot away. Yuk!

Question: What do astronauts do if their faces sweat and sweat gets into their eyes?

Answer: Honestly, there is very little we can do. Yet with a bit of ingenuity, we are not totally without some options.

Regarding tears and sweat in your eyes, the lack of gravity is a big deal. The sweat doesn't move or slide down your cheeks. Due to surface tension, it stays wherever it exits. But tears and sweat can be a problem due to their high salt content, which can cause an unwelcome sting.

Further, as part of our space walk prep, we are instructed to wipe the inside of our helmet visors with a specially prepared wipe and liquid antifogging substance. Similar to what you might use on a snow skiing trip and developed at the Johnson Space Center in Houston, this liquid's formulation is primarily made of the famous Johnson's No More Tears baby shampoo. I understand that before they used the No More Tears version, NASA used regular shampoo, which led to Canadian astronaut Chris Hadfield's TEDx Talk about how he "went blind during a space walk." A bit overdramatic, he could have simply said that he got soap in his eyes. Regardless, it would be somewhat debilitating, because you really can't do much about it in microgravity until your eyes tear up enough to begin to dilute the soap. The same can be said for sweat in your eyes, but it's much less painful. Blinking your eyes rapidly for an extended time can help speed up the generation of more tears, helping to dilute what was already there causing the issue.

Another option, which is not preferred by the ground team, is to open the helmet purge valve. Creating a small opening to the vacuum of space, the pressure differential between the inside of the helmet and outer space causes the liquid within to be sucked out quickly. The fear, though, is that the cold temperature of space may freeze the valve in the open position, allowing oxygen to leak continuously until the opening is closed.

All this reminiscing is making me tear up.

Question: Do astronauts experience a fear of heights during a space walk?

Answer: I was blessed to walk six times, totaling nearly forty hours in the vacuum of space and I never felt anything akin to what I had read or heard from other astronauts. A fear of falling, a need to grip too tightly to the handrails on the ISS, a sense of spinning, or a fear of heights? I felt none of it. Perhaps I was destined to be a spacewalker!

Question: Does the ISS create any gravity at all in respect to EVAs? (In other words, while space walking, does the ISS pull the astronaut toward the space station?)

Answer: Does the ISS pull the astronaut toward the station during a space walk? Did I paraphrase that correctly? If I did, the answer is yes but *not* for the reasons you think. The ISS creates no gravitational force to "pull" an astronaut closer as your question suggests. It's done with something mechanical, more akin to a fishing reel.

Every time an astronaut is required to leave the cozy and quite comfy interior of the ISS to perform a space walk, many precautions are taken to ensure that astronaut's safety. The suit is methodically checked and rechecked, as are all the tools and gear that will be used once our "handyman" is outside. First and foremost in that gear check will be the unit we astronauts refer to as our safety tether.

A tremendously clever piece of engineering, the safety tether consists of two hooks—one large, one small—connected by straps of space suit–like material on each end of a reel device that I liken to those used on fishing poles. This reel typically contains eighty-five feet of thin, braided cable wire, which can be extended automatically as the astronaut moves away from the ISS exit point known as the airlock hatch. (In the shuttle days—since the shuttle was so much smaller—these tethers were a mere fifty-five feet in length. We still use them on the station if needed.)

Having exited from the airlock opening, the first thing the spacewalker is required to do is attach the large hook of the safety tether to a specific tether point just outside the hatch. The smaller hook, which is affixed to the astronaut's space suit before suiting up and while still safely inside in the station, will remain in place *unless* the astronaut has to do a tether swap at some point during the EVA. While eighty-five feet of cable may seem more than long enough, the ISS is the size of an American football field or World Cup soccer pitch, so oftentimes we move to a point more than eighty-five feet away; then more than one tether is required. Hence, we perform a swap to a second tether. Think of Tarzan, swinging beneath his jungle canopy, moving from vine to vine to allow himself to move farther away from his original treetop. We just have to be a lot more careful during our swaps than he is!

Now these tether reels retract, and/or hold the cable, automatically—kind of like the seatbelts in our cars. And since the microgravity (weightlessness) of space doesn't need much force to pull you around, even slow, deliberate movements—of an amount of force larger than the cable's pull—let you move farther away while correspondingly extending the cable wire from the reel. The theory is, if you were to just let go, the reel's force would pull you back to wherever it is anchored, typically at the airlock if you are only using one tether. If you use multiple tethers during your space walk, the reel pulls you to the point where you swapped tethers.

Of course, additional, smaller, nonretractable tethers are also part of an astronaut's tool belt. These versions, referred to as local tethers, provide an additional level of protection above and beyond the safety tethers. As an astronaut moves to a new work site and is ready to stay there for a while, it is imperative that the local tethers are affixed immediately for redundancy and ensuring our hero won't be going anywhere, no matter what happens.

Question: How does it feel to do an untethered space walk?

Answer: Well, I don't know as I have never done an untethered space walk. To my knowledge, the first untethered spacewalker was Bruce McCandless. While using the manned-maneuvering unit, he sailed away from the space shuttle *Challenger* on mission STS-41B.

The SAFER jet pack—the current untethered flight provider—was first flown by astronauts Mark Lee and Carl Meade as a test flight on mission STS-64. It underwent further testing by astronauts Michael Lopez-Alegria and Jeff Wisoff on STS-92. Shown clearly in the photo taken during one of my space walks with the STS-118 crew in August 2007, SAFER has never been used for an emergency return to the ISS, its designed intent.

Question: What will happen to an astronaut if she or he gets detached from the ISS during an EVA? Will the astronaut drift away into space or fall back to the Earth?

Answer: Becoming detached from the ISS during an EVA is a low-probability occurrence. While not likely to happen, astronauts do prepare for it since it is possible.

For each and every space walk, one of the first and most critical steps occurs before ever departing the station's airlock. With hatches still closed and locked, astronauts verify their individual safety tethers are appropriately closed and locked on their space suits. In addition, while inside the station's "porch," the two spacewalkers (think buddy system) hook their safety tethers together in a move we call the daisy chain. Now both astronauts are connected to each other, and one of them is also anchored firmly inside the station.

25. While performing a space walk, the SAFER jet pack attached to the back of my suit is visible.

When the hatch finally opens, the astronauts execute another critical procedure. The lead spacewalker (EV1) will exit the airlock headfirst and attach the other, or free, end of his or her safety tether to an anchor point *outside*. Now the two daisy-chained astronauts are anchored both inside *and* outside the station. At this point, the EV1 will break the daisy chain and hook the safety tether of the second spacewalker (EV2)—who is still safely attached inside—to the external anchor point. Now both crew members are safely, and separately, anchored externally.

It is now time for the EV2 astronaut, who is still inside, to disconnect his or her internal anchor point and exit the hatch. Voilà! Two space-walkers, each safely anchored outside the ISS, are ready to go to work.

All of that effort may still not be enough if proper tether protocol is not exhibited by our brave spacefarers throughout the space walk. As they both move away from the airlock—with anchor points still firmly attached—each must endeavor to keep their individual eighty-five-foot lines separate and untangled, a task that seems simple but isn't always so, during a six-and-a-half-hour excursion outside in a microgravity environment. In addition, if the work site is more than eighty-five feet away from the starting point, additional safety tethers must be carried with them, and tether swaps must be appropriately and safely executed as needed.

Yet I haven't really answered your question, have I? If all of our tether protocol efforts still fail us—remember, this is an extremely low probability—and one of us totally detaches from the ISS structure, we will be ready, albeit a tad bit embarrassed.

While it's impossible to fall all the way to Earth, as your question suggests, it is possible the rates (e.g., of departure speed and rotation) imparted to a spacewalker who has lost physical contact with the station could be high. The resulting drift away from the ISS and deeper into outer space would not be good. Rescue by another spacecraft would be futile, as it would take too long to undock, rendezvous, and capture the wayward astronaut. Another option was required.

The solution? It's pretty amazing. Like Buck Rogers, we have a jet pack attached to our suits, but *unlike* Mr. Rogers, we have virtual reality to train for this exact scenario.

The training venue is called the VR Lab (original, huh?). Home to the true geniuses behind the Dynamic Onboard Ubiquitous Graphics software system (nicknamed DOUG), we are trained to utilize our last resort—the SAFER backpack—in a virtual world.

Wearing special gloves and helmets helps create a virtual world that enables us to have a simulated fall from the station, and repetition is key to a safe return. Since the SAFER is our last resort, we must be ready to correctly deploy its hand controller and fly ourselves back to the station. "Fly" is a general term. In essence, we must learn to understand and use the nuances of orbital mechanics to get our EVA suit back to the ISS spaceship. Practice makes perfect. We see the ISS virtually, and our inputs to the hand controller, coupled with the software's execution, give us a video game of sorts that imitates what we would experience in space. As our ability to return safely progresses in the virtual world, we fall off with higher and higher rates until we can consistently demonstrate that we can save ourselves repeatedly.

It was *fun* to be an astronaut!

Question: What happens when astronauts are out of the oxygen in the tank? Because, according to the movie *Gravity*, it says, "You are out of oxygen in tank but have it in suit." What does that imply?

Answer: I think it means you're gonna die! An astronaut's space suit has two supplies of oxygen. One is the nominal/regular supply, which is typically good for a six- to eight-hour space walk assuming normal breathing rates. A thirty-minute backup, or secondary, oxygen supply is also self-contained in the suit for use in an emergency that requires an EVA abort. An "abort" means we must return to the airlock within that thirty minutes or risk having no more oxygen at all. This is *not good*, and we practice for this in NASA's Neutral Buoyancy Lab in Houston.

The movie may refer to the fact that even if your main tank is empty, you still have thirty minutes in the secondary. *Or* it could mean that if your suit tank O_2 supply is exhausted, there is still a very small amount that could remain within the lines of the suit; and that ain't much! You'll have to ask the movie's director/producer.

Often astronauts return to the airlock to replenish their suit's main O_2 tank. Usually this is because they are working far too hard and breathing far too fast, thereby depleting their supply too quickly. This can happen on a rookie's first space walk or if a difficulty pops up and requires the astronaut to exert much more effort than normally required. As mentioned, my rookie space walk lasted nearly eight hours, and I had so much O₂ left that we could have stayed out for another ninety minutes or so. Guess that means I was born to perform space walks.

Question: Do astronauts use their legs or feet during space walks? Why do space suits have shoes?

Answer: Sometimes we *must* use our feet. Using our feet, by fixing them in what we call an articulating portable foot restraint, allows us the freedom to use both of our hands. For many spacewalking tasks, that is an extremely important consideration. If you are near certain parts of the ISS and your feet are of the right (smaller) size, you can actually wedge your space boots into part of the truss structure and kind of give yourself a foot plate for "free." Other than those instances, feet are not really necessary for a space walk. Perhaps that's why Kevin Michael Connolly of the Travel Channel's *Armed and Ready* television show did so well when he participated on one of the Johnson Space Center's spacewalking simulators called the Active Response Gravity Offload System.

Space suits must have boots to help us seal the pressure vessel that is our space suit/spaceship. We wear those suits to protect us from the vacuum of outer space, and it's much easier to make the suit a sort of body glove or cover so we can guarantee the integrity of the various connection joints: boots, gloves, helmet, and so on. Happy spacewalking!

Question: How do astronauts perform tasks outside of the ISS when it's moving at 17,500 miles per hour? How do they even get out of the ISS? Would it be the same as jumping out of a car traveling 70 miles per hour on the highway and be left behind immediately? Not to mention that the speed is 250 times greater. Why do they call it a space walk when you are literally flying faster than Superman?

Answer: The answer to your questions is that it's all relative—relative motion, that is.

26. Exiting the Quest air lock, ready for the start of another space walk.

During a space walk, it's true the ISS is moving at 17,500 miles per hour about the Earth. But the spacewalker, who crawls from within the ISS, is also traveling at 17,500 miles per hour. Relative to one another, they are—for all practical purposes—not moving much.

In your question you ask if it would be the same as jumping from a car on the freeway. If spacewalking astronauts jumped from the ISS and were not tethered to it per the normal protocol, they too would be moving relative to the ISS. Depending on the direction and the speed they departed, their separation distance would probably increase—that is, sort of similar to "getting left behind immediately"—in whatever direction they jumped. But the physics of outer space, or orbital mechanics, is a bit different than on Earth. It's possible that, if left alone, the leaping-from-the-ISS spacewalkers would return to nearly the same point from which they departed one orbit later. That's orbital mechanics for you.

But our friend who leaped from the car will experience wind resistance, gravity, and other variables that contribute to the resultant motion for him or her relative to the car speeding down the highway. In space, gravity's effect is much, much less; there is no wind resistance; and so on. It's a slightly different problem. If our friend carefully crawls from the car window and slowly moves around the outside of the car, then our situation is more akin to that of a spacewalking astronaut. Confused? Me too!

As noted, the official term for these forays into the abyss of outer space is "extravehicular activity," or activity outside a vehicle. Why it morphed into "space walk" I'm not sure, but while on the Moon, the astronauts did walk, just as they will one day walk on Mars.

I'm no genius, and there are those out there much smarter than I (that's your cue, Robert Frost) who can perhaps provide a much more technical answer to this wonderful question. But for now, if you're going to jump from your car or the space station, make sure you're tethered to something, and keep lookin' up!

10

So You Wanna Be an Astronaut?

Question: If you're in the International Space Station, do you feel warmth from the sunlight that passes through its windows?

Answer: Absolutely you can feel the warmth from the sunlight passing through the station's windows! One of my favorite things to do while living on board the ISS was to fly into the Russian segment—at that time, only the Russian modules had windows looking to the sides and, thus, at the sun—and put my face up against the window. As the sun would rise from its half-orbit concealment behind our Earth, I could feel its warmth on my cheeks and its bright light bathing my eyes.

To gain the maximum effect, I would keep my eyes tightly closed and imagine that I was back home on Earth, lying with my family in our backyard on a beautiful summer's day and soaking in those life-giving rays from our solar system's star some 93 million miles away. For just a few short moments, I was "back home," and it felt good. It felt really good!

Question: Are you and Robert Frost friends? You mentioned that you are starting to piggyback off of Robert Frost's technical answers more frequently. Then you mention the two of you as a "tag team." Have you met each other in real life? If not, what are your impressions of each other?

Answer: Are Robert Frost and I friends? I would certainly hope that he feels the same way about me as I do toward him. Robert and I first met when he was a young instructor in the United Space Alliance group within the Mission Operations Directorate that taught "baby astronauts" the nuances of the ISS motion control system (MCS). A somewhat complex system that helped the ISS maintain attitude control in space only after being integrated with the analogous Russian segment version, Robert and his cohorts were able to make the training both efficient *and* inter-esting by using innovative concepts such as props—with astronauts performing in the roles of interacting control moment gyros—and a parody of the game of *Jeopardy!*, where crews were challenged by ques-tions and answers geared at reinforcing key system concepts. By the way, our Expedition 15 crew held the record for points scored (for a time anyway) based on all of our correct answers.

The MCS group probably remembers me as the astronaut who reminded them with every training session that "I hated their system." Trying to be playful, as always, I used that line to tease them about the complexities that they were making so easy to understand. A few of them, however—and I hope Robert was not one of them—were unable to understand my pokes as playful fun and took them more as virtual "barbs" into their young and pliable "MCS flesh." They couldn't have been more wrong.

I would go on to fly two missions in space—a 152-day mission to the ISS and a 15-day resupply to that very same laboratory complex. I would not have been successful at all without the help of people just like Robert. I find him to be extremely intelligent and easy to talk to and a wonderful instructor. Like so many other astronauts, I was fortunate to be learning from his expertise. Thank you, Robert Frost. I owe you!

Robert Frost says,

Clay and I met 15 years ago when he was a baby AsCan (Astronaut Candidate). His class (the 17th AsCan class) was huge. If I remem-ber correctly, it was 32 people, including a few JAXA [Japanese Space Agency] and ESA [European Space Agency] students. Their nickname was the Penguins because there were so many and ISS was still a fingers-crossed dream. Penguins, as you probably know, don't fly.

Look at those baby faces! Swanny ([Steve Swanson,] standing next to Clay) just returned from ISS [2014]. His hair is a bit grayer, today. That class was my favorite AsCan class. They were also the ones I got to teach the most, because since ISS wasn't yet on orbit, we didn't yet have a firm grip on what the crew really needed to know, so we taught them everything.

While Clay was a new AsCan, he wasn't new to JSC, having previ-ously worked in DM (flight design and dynamics). After AsCan training,

27. The Penguins, the seventeenth group of U.S. astronauts and our international colleagues, pose for an early career photograph, 1998.

astronauts receive technical assignments while they await flight assignments. They also help out in simulations, playing the on orbit crew.

Those activities and the fact that we're all in the same building means we frequently pass each other in the halls. Astronauts take hundreds of classes and meet thousands of people, so it always feels a little special when you pass in the halls and they greet you by name. Clay is a product of the Midwest. He always stopped to shake hands and greet me by name and usually added a few extra words to the effect of *"Hey Robert! How's it going? I hate MCG!"* MCG (Motion Control Group) was the system I used to teach. It was a bane to astronauts because it wasn't really a hands-on system for them, as long as it was working. But a loss of attitude control could quickly ruin their day and recovering from such a scenario required understanding the complicated integration between the U.S. and Russian segments.

Most kept their disdain for the system to themselves. Clay openly (but kiddingly) disdained us. Only his classmate Tracy Caldwell routinely greeted us with *"I love MCG!"* (she actually called me one day to ask to take the final exam again, just for fun).

I had the privilege of accompanying Clay, Sunita Williams, and Butch Wilmore . . . to Cologne, Germany to observe their ATV training. Clay and Suni were kind enough to ask me to join them downtown to visit the Dom (cathedral) and have dinner. I'm a history nut, so I think I bored them to tears lecturing about the Roman ruins under the Dom.

It was a pleasure working with him at NASA and it's fun tag-teaming on questions. . . . It was touching reading his answer to this question.

Question: How many astronaut jobs are out there?

Answer: Not many astronaut jobs are "out there" in the United States. In the most recent astronaut selection cycle (2017), NASA chose twelve U.S. astronauts (and two Canadians) from more than eighteen thousand applicants!

As noted previously, when I was selected in 1998, twenty-five American astronauts were selected from more than twenty-six hundred applications, or roughly less than 1 percent. Throw in seven international "baby astros," and our total was thirty-two. The 1996 class topped out at forty-four. Today with no capability to launch from U.S. soil, our seats on board the three-person Russian Soyuz spacecraft are aggressively pursued by our partner nations and their agencies—JAXA, ESA, and so on—which are steadily gaining traction in their ability to fill those seats. I am hoping the commercial spaceflight companies of Boeing and SpaceX will alleviate that concern and *soon*.

Today, there are about 44 active (eligible for flight) American astronauts in the corps, way down from my initial years when the number topped 100 (it would reach 149 in the year 2000). With a U.S. governmental budget of less than half a penny going to our space program, something has to give. I guess it's the number of astronauts.

Regardless of these sobering statistics, *apply!* When NASA is looking for astronauts, *apply!* Go to the National Aeronautics and Space Administration website (www.nasa.gov) and see what they are looking for. Then let NASA tell you whether you are—or are not—"selectable." Don't make that determination yourself. If you have physical issues that are not precluded on the NASA application but they have you doubting your candidacy, apply anyway. NASA will let you know; trust me on that one. As a guy who applied fifteen times to secure my dream job of a lifetime, I am here to tell you that you can *never* give up. For me, persistence paid off.

Question: How does an astronaut feel after returning from outer space?

Answer: My second return to Earth, coming home on *Discovery* with the crew of STS-131, was much better than my first. With my balance a bit shaky at first, I quickly recovered enough to exit the orbiter and perform the walk around with my crewmates. I would take a two-hour nap upon reaching our condominium and then join our crew families and friends outside near the pool to reminisce about our experiences.

In my first book, it took a full chapter (17) to cover this question. Some highlights are provided here.

> Safely back on the ground, I was in hog heaven, contemplating seeing my wife and kids after five long months.
>
> [Kennedy Space Center] KSC's astronaut support personnel opened the shuttle's side hatch, and veteran astronaut Jerry Ross poked his head in.
>
> "Welcome Home!" he said with a huge smile.
>
> An orange corrugated hose, or "elephant trunk," was inserted through the hatch to pump cool air into the rapidly warming middeck. I could smell the scents of Earth, and it didn't even matter which smells they were. It was awesome to be home. At least it seemed that way from my comfortable vantage point of lying on my back.
>
> Jerry coordinated the extraction of each crewmember in a clear and specific order. The "ISS guy" would be last.
>
> Scott [Parazynski] and Wheels [Douglas Wheelock] exited their middeck seats quickly and apparently readapted to Earth's gravity within minutes. They hurriedly departed *Discovery* in anticipation of doffing the bulky, heavy, extremely warm suits they had been perspiring in for the last hour.
>
> Turning my head ever so gently to the left (now that I was back in the firm grip of Newton's second law, moving too fast would cause

"stomach awareness"), I watched as my crewmates were brought down the ladder from the flight deck, starting with my good friend from Italy, Paolo Nespoli.

As Paolo slowly descended the short metal ladder under the watchful eye of Ross, I shouted, "Way to go, Paolo! Great job!"

He slowly turned his head toward me, and, with great effort, uttered a quiet and unconvincing "Thanks." That was followed by the audible splat of a discharge of fluid from his stomach onto the middeck floor.

Daunted by this resonant and disconcertingly visual display of fluid unloading, I returned my head to neutral and concentrated on the switches and dials mounted on the ceiling in an attempt to get that puking out of my mind and to once again bask in the sense of complete success that had previously washed over me.

Finally it was my turn. Jerry Ross placed a calming hand on my left knee.

"Are you ready?" he asked.

"You bet!" I said, having no idea whether I was or not.

He undid my remaining parachute-to-harness straps. (I had already released my five-point seatbelt harness, as it's a simple device requiring only the turn of a knob.)

Watching for the telltale signs of uneasiness that only a veteran space flier could see, Jerry asked me to slowly sit up.

It took me a great effort to rise into the seated position, even with an assist from Jerry's strong arms. Upright for the first time on Earth in over five months, the entire middeck of the orbiter began to spin counterclockwise at an incredible rate. Fighting off nausea, I focused on one of the cream-colored lockers we'd opened and closed a hundred times as we packed up. The locker seemed to stare right back at me. Keeping my gaze affixed on my newly found reference point, I resisted the urge to turn my head as the shuttle support team shouted instructions to each other as they began to unload our personal gear.

It took only a few seconds for the spinning of the middeck to slow down and ultimately come to rest, presenting me with the view that any earthbound astronaut would have expected.

Jerry Ross asked again, "Are you okay?"

"Yes," I responded without much enthusiasm.

"I need you to turn to your left and get down on the floor. You will have to crawl to the hatch," he said.

I turned left, ever so slowly, anticipating the moves needed to get myself to my hands and knees on the shuttle's middeck floor. I took a deep breath. Positioning my hands forward to catch myself should I lose control and fall in a pumpkin-orange heap, I made the move. The thickness of my launch-and-entry suit protected my kneecaps from the hard floor after 152 days of treading only on air. Success was achieved until I moved my head to look at the open hatch and freedom.

The spinning started. Once again, I held firm. My head remained as stable as the faces on Mount Rushmore. It took less time for the spinning to stop than before. Confidence washed over my tired and overheated body. With newfound vigor and the hope that I was going to be able to exit without puking my brains out, I allowed myself the fantasy that I might even be able to perform the shuttle walkaround with the rest of the crew.

It took a considerable amount of strength in my arms and upper body to pull my two-hundred-pound self into the opened hatchway. Nearly exhausted from the effort that took only seconds, I was greeted by two able-bodied flight surgeons. I gave them a weak but sincere smile as they hoisted my arms around their shoulders and lifted me from what was essentially a prone position. It was time to try and walk again. For the first time in over five months, my legs began to receive commands from my brain, the orders flying at the speed of light through a nerve system that seemed to be relearning everything from scratch now that gravity had returned. As if I were Tim Conway playing Mr. Tudball on *The Carol Burnett Show*, my size-thirteen black flight boots shuffled slowly across the gantry way to the door of the crew transfer vehicle (CTV).

My intestines were having even more difficulty making the transition to normalcy. I had gone from mostly Russian food on the space station to a diet of American food aboard the shuttle. Thus, my internal organs had been in a constant state of gaseous protest for the last two days of the mission.

Question: Is it true that NASA continued to use 1960s and 1970s software technology during the space shuttle era into the 2000s because the software was bulletproof (error free)?

Answer: My understanding of the software is that NASA continued to use it not because it was bulletproof but because it was well understood. Over the years, this special code, which was developed specifically for the space shuttle program by contractors, continued to be validated—line by line, subroutine by subroutine—from constant simulations, testing, and upgrades. The decision to continue its use, rather than upgrade the code to faster and more stable coding languages, was born of the fact that it was becoming better understood with every passing simulation and with fewer possibilities for uncovering a new error or problem in the code. I guess in some sense that can be said to be bulletproof, but the code was edited in some fashion almost until the end of the program.

Question: How does it feel when you celebrate someone's birthday in space? Has anyone celebrated their birthday in space?

Answer: I did not celebrate one of my own birthdays while I lived in outer space. However, I did celebrate the birthday of my wife.

The date was July 2007. My wife's birthday was fast approaching, and I knew that I needed to do something special to honor the occasion. And not being a resident of the planet Earth at the time, it would take a bit of extra effort.

First, I needed to remotely secure her birthday present. With the help of email (the ISS had no internet connection back then) and a good friend on the ground, I was able to purchase jewelry that would complement some of her existing paraphernalia! The gift was scheduled to be delivered to her, along with flowers, on her birthday.

Next on the list was a way to surprise her and communicate to the entire world that it was her birthday. After all, I was in outer space. For this, I needed help from the Mission Control Center and the CAPCOM. Now an astronaut, Mark Vande Hei was the lead CAPCOM at the MCC for the ISS Expedition 15 mission and agreed to help me with my plan.

My wife was employed at the Johnson Space Center in Houston, at that time as a public affairs education specialist. I had learned through our email and phone conversations that she would be providing a tour of the JSC's Space Vehicle Mock-up Facility (Building 9) for a large group of teachers on the morning of her birthday. Perfect, I thought! It was time to put my plan into action.

By conversing with Vande Hei, I was able to determine a rough time as to when my wife and her group would arrive in Building 9. The operations control desk there has a digital TV set up to provide visitors with NASA TV whenever it is live from the ISS. This would be key in my plan. With ISS on Zulu time, or Greenwich mean time, it would be later in my day when the tour group would be showing up at operations control around 10:00 a.m. their time. With my personal "spies" placed throughout the agency, the whereabouts of my wife and her group were being relayed to Vande Hei periodically throughout the morning.

Meanwhile, on board the ISS, I was ready. Keyed up on my station support computer was a waveform audio file format of what my wife and I referred to—early in our dating years—as our song. Recorded by Bryan Adams as the theme song for Kevin Costner's *Robin Hood*, "(Everything I Do) I Do It for You" was the song we both associated with our courtship and marriage. Also available in my home away from home were five multicolored balloons that were a bit low on air but still a visual reminder of some past ISS celebration. I maneuvered them to an appropriate spot that would be clearly visible in the laboratory module's aft-facing video camera.

As we approached "time zero," when live ISS video would be broadcast to the world via NASA TV, the speakers for the ISS communications system came to life. The call was from on-console CAPCOM Vande Hei, measured in tone, and appropriate for the army officer he was: "Station, this is Houston on Space-to-ground 2. The Penguin is on the ice. Repeat. The Penguin is on the ice." Not normal parlance for our regular up and down communications, this was our covert signal that all was ready on the ground. My wife and her tour group were mingling in Building 9, grouped in front of the operations control desk and its high-definition television screen. It was now or never.

With a quick push of the laptop's Enter key, I moved the now-keyed ISS communications system's microphone in front of the computer's integral speakers. With me floating in zero gravity, clearly in front of a now active and live-to-the-world video image, our song began to play. With thirty-plus school educators watching and listening intently, my wife's birthday surprise was being broadcast to Earth from outer space! It was a birthday she would never forget, and the tears began to flow. With Susan caught in the adulation of folks she didn't even know, I grabbed the mic at the song's completion. With emotion mustering up inside me as well, I told the world of my unending love for the beautiful and talented woman I had been blessed to call my wife and, in closing, wished her a very happy birthday.

As my brother would say later, "Most men in America will be mad at you for setting the bar so damned high." That's okay. My wife deserved it.

Question: What would happen if an astronaut blew air into a musical instrument in outer space?

Answer: In the vacuum of outer space, where there is no air to carry the sound waves, I guess nothing would happen. And with having to remove my helmet being a key step in the process, I don't really want to be the person that tries to find out. Inside the ISS, where there is an atmosphere just like on Earth, one could easily hear the instrument.

Question: Can you be upside down in space?

Answer: Heck yeah! Check out the photo from my time in outer space on mission STS-131.
And, as they say, a picture is worth a thousand words.

Question: Why can't I see the live footage from the ISS on NASA's website?

Answer: If your computer has the appropriate software and drivers loaded (and is plugged in and powered), you should be able to see NASA TV via the website. Good luck, and keep lookin' up and at NASA TV! Note that there is not always live footage available throughout the day.

28. It *is* possible to be upside down in space. Actually, you can be in any position you want as I demonstrate with *Discovery*'s pilot and my STS-131 crew mate Jim "Mash" Dutton!

Question: Is it difficult to swallow in zero gravity?

Answer: Nope, not at all. I lived in outer space for five months, and I swallowed in zero gravity just as you do here on Earth. No differences that I can recall. It *was* hard for me to choke down chicken in pesto sauce, but I don't think that had anything to do with zero gravity.

Question: Is it possible to create a bubble in zero gravity?

Answer: Take a look at the accompanying photo and see for yourself! It all happens because of physics. In a microgravity environment, the liquid's surface tension dominates, making it take on the default shape of a perfect sphere.

Question: How is the interior cleanliness maintained aboard the International Space Station? Curious about how the inhabited interior is

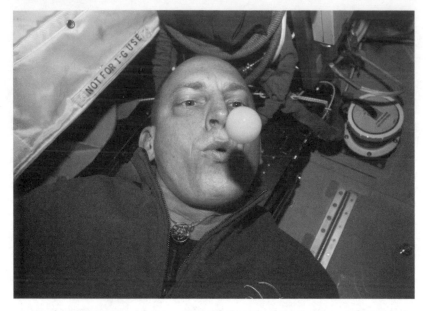

29. Peach-mango juice makes a very visible liquid sphere in *Discovery*'s middeck on STS-131.

maintained in general, in light of the sensitive equipment that may get clogged. I had in mind astronauts shedding hair, skin flakes, and so on.

Answer: If I answer this question, please don't tell my wife! You see, on board the ISS we were scheduled to clean our home every Saturday morning. Since I was the only American astronaut living on the ISS at that time (June–November 2007), I was responsible for cleaning the entire U.S. segment. If she finds out I actually *do* cleaning, it could spell doom for me here in my own house in Houston!

While the station's systems keep things reasonably clean automatically by sucking up all types of nastiness into awaiting high-efficiency particulate air filters, astronauts clean the ISS for between two and four hours every Saturday morning per the schedule. I have to be brutally honest here and tell you that I did it in *much less* time than that and mostly because I didn't do the myriad tasks the ground requested.

For example, every Saturday we were to wipe down all handrails and communication panels, or audio-telecommunication units, on the ISS and a bunch of other stuff. I wiped them down but not every Saturday. I only did it right before or right after a visiting crew was coming or had been aboard. My reasons were—at least in my mind—quite valid. You see, disinfectant wipes are a consumable; as such, you can use them up. Consumables cost money, and the more you use the more frequently they must be replenished through a cargo ship delivery to space. With cargo mass being quite precious—not to mention expensive—I figured I was helping the program save money by reducing the consumption of consumables such as disinfectant wipes and thus increasing the ability to send, say, more important things like food containers or clothing or toilet wipes, for example. Also, it made no sense to me to disinfect those things when ten minutes later, Oleg or Fyodor or I would sail through the ISS as always, grabbing handrails and audio-telecommunication units, putting our grubby little hands all over everything. So I treated the Saturday morning cleaning ritual much as I might here on Earth. Before and after visitors, I gave the place a thorough cleaning. But most Saturdays, it was a less focused effort. I'm sure after providing this answer, the folks in Mission Control will be pretty pissed, but no one died or got sick because of it.

Vacuuming filters and vents I took more seriously. These places actually accumulated much gunk, leading to the possibility of reduced airflow in our very important ventilation system. Plus, it was easy to see progress as the powerful suction of the ISS vacuum cleaner provided instant gratification for my efforts. It was quite satisfying for me to see lint, dust, and twelve- to fifteen-inch strands of Suni Williams's, Barb Morgan's, and Tracy Caldwell-Dyson's hair disappear into the vacuum's bag (also a consumable). I would vacuum up everyone's hair (but mostly theirs!) every Saturday for the entire five months I lived aboard the ISS.

The key to living, working, and surviving in outer space is to take care of your stuff. If you and your crewmates are diligent and form good habits about taking care of the things that are your responsibility, all will be well. I remember arriving back at the ISS with the crew of STS-131 and observing the state of cleanliness and organization of the ISS Expedition 22/23 crews. I told T. J. Creamer that if he'd give me about an hour, I could have the place looking shipshape. I don't think he appreciated that too much.

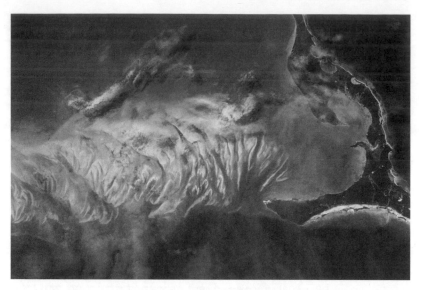

30. The island of Eleuthera and Rock Sound in the Caribbean Sea.

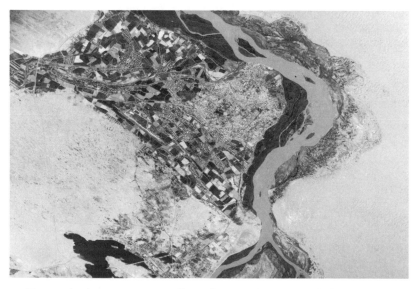

31. The nearby Amu Darya River allows the city of Termez, Uzbekistan, to flourish through irrigation.

Question: What are some of your best pictures on the ISS and in space?

Answer: I have included a few of my favorite—not sure if that means "best"—pictures from inside the space station and in outer space. Enjoy!

Question: Does a person in space need less food to survive than he or she would need on Earth?

Answer: I don't really know the technically correct answer to this question. I'm not a nutritionist. However, what I can tell you about is what I experienced during my stints on board the ISS in 2007 and 2010.

Fortunately, I got nowhere close to worrying about survival due to food intake or lack thereof. As a matter of fact, even though we were targeted to consume a certain level of calories (or less), we pretty much had free rein during our mealtimes. My Russian crewmates Yurchikin and Kotov and I ate three squares a day and often had snacks during our twice-daily coffee breaks. Usually, I ate like a horse. My intake was even higher after tough workout days or after performing a space walk, which could last anywhere from five to eight hours. There were a few days when, due to stress or not feeling quite "up to snuff," I would eat a bit less than usual.

During my five-month mission I lost twelve pounds. Upon returning to the ISS on the space shuttle *Discovery* with the crew of STS-131, I also lost twelve pounds. My uneducated hypothesis is that when your body reaches a unique environment such as microgravity, its immediate response is to rid the body of what it doesn't really need. Apparently, in my case, it needed to be rid of twelve pounds.

Our bodies can survive much easier on a reduced food intake than a reduced water intake. That is well documented. In space, we had ample supplies of both, thank goodness.

Question: While in space would astronauts be discouraged from consuming coffee? I'm thinking about this over my morning coffee. I'm asking because coffee is a diuretic and contributes to dehydration.

32. As a basketball official in zero gravity, I could fly up and down the court!

33. Peanut butter and honey on flatbread in zero gravity!

Answer: As far as I can recall, I was never discouraged by anyone from drinking coffee while living on board the ISS. For me, morning coffee was one of those wonderful psychological benefits, reminding me of life here on Earth.

I limited my coffee intake to two "cups" per day. I say two cups, but it is actually two bags of coffee on the ISS. While typically calling for eight to ten ounces of hot water to mix with the freeze-dried coffee within, I usually upped the amount of water to twelve to fourteen ounces, thereby giving me more coffee to drink and perhaps a less caffeinated version. My preferred coffee was Hawaiian Kona coffee with creamer and sugar. If I didn't have this available to me, I didn't drink any coffee at all. Yes, I am apparently a coffee snob!

My crewmates, Station Commander Yurchikin and Flight Engineer Kotov, both *loved* coffee, as do many Russians. Fyodor was a "coffee black" kind of guy, while Oleg liked his black with sugar. Seemed our tastes in the coffee bean nectar category fit together perfectly. However, I learned early in our five-month stay that they both drank *way* more coffee than I did, and once their individual stashes were depleted, they went after mine! It wasn't until I called the Mission Control Center and requested more of my personal coffee choice be sent up that the STS-118 crew— led by the soon-to-fly-for-a-full-year commander Scott Kelly—brought us an ample new supply in August 2007. After that point, I had plenty of coffee for the remainder of my stay in orbit.

My final coffee story came about in the Russian service module. Enjoying my morning breakfast and prepping a fresh bag of hot coffee, I had attached the silver bag to the Russian water processing system that pumps the hot water into the bag. Counting the "hum" of the pump's cycles—our way to easily measure the amount of water entering the bag—all was going well until Oleg asked me to help him on the floor of the module. Thinking it would only take a second or two, I left the bag as it kept filling. Seconds growing quickly, the bag continued to fill until it burst at the seam, sending scalding hot coffee with creamer and sugar flying onto the surface of Oleg's bedroom door.

With a scream and maybe a cuss word or two, I flew to the perfectly performing system and closed the appropriate valve, stopping the flow of water into the bag and onto the carpet-like covered door! While I was feeling horribly defeated and embarrassed, Oleg told me in near-perfect English, "Do not worry, Clay. It has happened before. Fyodor did it too!" With a hearty laugh, not feeling quite so bad anymore, I cleaned up the mess, and life continued with the crew of Expedition 15.

Question: Could an astronaut play football in space?

Answer: You can adapt and play several sports in the microgravity environment of the ISS. Even in an environment lacking gravity, some ingenuity and care let us play Nerf basketball—we called the game "ISS" instead of "PIG"—baseball, soccer, and American football. As a matter of fact, I kicked field goals in the space station's lab module, aiming at the video camera mounted on the forward end. Guessing the ground control team didn't like seeing footballs flying at them from space! And, of course, if we're suiting up to go outside, wouldn't our spacewalking helmets need the appropriate American football adornments—in my case, a big, red *N* for Nebraska? Go, Big Red!

Question: Is time spent in space logged in an airplane-style logbook? How about the time spent in the atmosphere on the way there?

Answer: No, we don't really log our time in space in an airplane style logbook, but our time in space is logged by the good folks on the ground. When I returned to Earth following my stint as a flight engineer on board the International Space Station, my lead flight director, Robert Dempsey—also known as "Dr. Astronomy"—was ready to provide me with my on-orbit duration. He casually told me I had been in space 151 days, eighteen hours, twenty-three minutes, and fourteen seconds. Among friends we round up—astronauts like higher numbers—to 152 days. He did point out that he gave me credit for the time it took the shuttle *Atlantis* to reach the unofficial "astronaut international" altitude limit of 100 kilometers, or 62.1 nautical miles, which is measured from the sea-level anchor of the Kennedy Space Center's launch pad 39A to the imaginary threshold at our atmosphere's edge. I am assuming this is standard practice when NASA determines an astronaut's in-space flight times.

While flying the T-38, many of the instructor pilots would offer to sign the civilian general aviation logbooks for those of us who had our flight certificates, thus allowing us to document our NASA-modified U.S. Air Force trainer flight time for future reference. I never took them up on it, and I'm not sure how the hours were logged by my astronaut colleagues who served in the military. To me, it seemed too difficult a process with regard to filling out Federal Aviation Administration paperwork in hopes—there were no guarantees—of receiving actual credit for the flight hours. I assume, though, that some of my colleagues did. Some of my Penguin classmates were certified flight instructors and instrument rated, and perhaps those hours helped them achieve future flying goals.

I received a printout from the meticulous and now-retired Mavis Ilkenhans, who tracked all of this data electronically, when I retired in January 2013. The printout showed I had the following times:

Simulator hours: 6.6 (T-38)
Actual instrument hours: 828.6 (T-38 and general aviation aircraft)
Simulated instrument hours: 7.7 (T-38)
Night hours: 741.9 (all)
Total flight hours: 1,469.0

These hours included my time as a mission specialist/copilot in the KC-135 and Boeing 727 (zero-g airplanes), the shuttle training aircraft, and the space shuttle orbiters (*Atlantis* and *Discovery*), and the first-pilot time in three general aviation aircraft: Cessna 172, Cessna 150, and Piper Archer-28. I also got credit for 0.1 hours of first-pilot time in the T-38 when I was allowed to perform a touch and go from the backseat. Awesome!

I would end up logging about 167 days in outer space, being grateful for two forays above our atmosphere and nearly forty hours of spacewalking time spread over six space walks. Those numbers work just fine for me. They're easy to remember, and they make me very proud.

Question: Does the International Space Station have internet access? Do the flight engineers have direct access to the internet, or do they always have to liaise with ground control for Twitter posts, Foursquare check-ins, etc.?

Answer: Since early in 2010, the ISS has had internet access. Originally feared to be an easy pathway for internet hackers into our high-tech data systems, NASA has since been able to figure out how to provide ample cybersecurity.

As noted in chapter 8, when I visited the ISS as a member of the STS-131 *Discovery* crew, I tried to send some tweets to Earth using that internet system. I used JAXA astronaut Soichi Noguchi's laptop computer, but it was so extremely slow that I gave up after fifteen minutes and resorted to the system I had been using on the shuttle. That meant I typed my 140-character tweet and emailed it to the ground, where my contact in the Public Affairs Office posted the tweet to my Twitter account (@ Astro_Clay). Not efficient, but effective and probably just as fast as using the space station's internet.

Today, due to technology improvements in bandwidth and such, the internet system on board is a bit faster. For example, it is my understanding that the astronauts on board who use Twitter and other social media platforms are able to shoot a photo and tweet it moments later. Ah, technology. Probably would've been much easier to buy my wife flowers on our anniversary.

Question: How do I become an astronaut? I am a twenty-six-year-old software developer working in a good IT firm. Is there a way I can become an astronaut?

Answer: The best way to become an astronaut is to follow your passion. To find more information about qualifications, it is best to peruse the appropriate U.S. government websites—www.nasa.gov or www .usajobs.gov.

Question: Do you have some favorite pictures taken inside the space shuttle while in space?

Answer: I certainly do! I don't know if you will consider them good, but I have included a couple of shots that were taken on board *Discovery* during our fifteen-day mission, STS-131, in April 2010. Enjoy, "live long, and prosper." By the way, I was the photographer!

34. My crew notebook floats weightlessly on *Discovery*'s flight deck during a burn of its orbital maneuvering engines on STS-131.

Question: Are astronauts' families still given a "squawk box" (a speaker to play communications with astronauts) while their family members are in space? During the early days of the space program—Gemini and Apollo, but not sure about Mercury—the families were given a squawk box so they could hear what was happening during the mission. Does this still happen, or are there more modern ways this is achieved?

Answer: A squawk box is no longer provided, but NASA TV provides near-constant coverage. In addition, with today's space-age technologies, families of American astronauts aboard the ISS are loaned all necessary equipment for conducting weekly teleconferences from space.

Prior to my mission, a local internet/communications provider came to our home and installed everything we needed for our future video chats. It is amazing what we can do today with technology, and it made a tremendous difference for me to be able to see and talk with my wife and kids. Later, when the kids got bored with seeing Dad from space, we invited close friends and family over so that they too could enjoy the experience. Those were some of my most memorable moments.

35. Imitating Commander Spock from *Star Trek* on *Discovery*'s middeck.

I heard barbershop quartet singers (thanks, Pete Hasbrook), talked with Melinda Gates and her son, chatted with Olympic softball player Jennie Finch, and traded drumsticks with Chip Davis from Mannheim Steamroller and jokes with Yakov Smirnoff ("America, what a great country"). Yakov's best line was, "It is sad when American astronaut is funnier than Russian comedian!"

But no squawk boxes.

Question: Do ISS astronauts have habits in order to always be oriented in the same direction? Are modules of the space station built without a floor-ceiling relationship? Do astronauts have a preferred position relative to the Earth to work and sleep—with their feet pointing toward Earth, for example? If there are two astronauts in the same module, do they tend to orient themselves in the same direction, or can they work independently?

Answer: I think some of you spend way too much time reading stuff on the internet. Here is my answer to questions like this. The station's modules are indeed designed with clear definitions and orientations for the floor and ceiling. The expectation is that we will operate in a normal, earthlike fashion, with our feet near the floor and our heads near the ceiling. Lights are placed above our heads. Electrical outlets are installed near the floor.

As noted in chapter 8, on the ISS the U.S. segment's modules even have a "location coding" system that helps orient astronauts. Labels and decals abound, letting astronauts know or reminding them of the expected orientations and that the floor and ceilings are usable volumes. All station modules, including the international ones, have a defined floor and ceiling that are known respectively as "deck" and "overhead," and right and left as "starboard" and "port." If we move to the front of the ISS we are heading "forward"; flying back to the Russian end of the ISS is known as "aft." Very military, you say? That's because much of our heritage comes from the military. Learning these standard orientations makes it easier for space fliers from all countries to adapt and understand everything from stowage locations to emergency egress, or exit, instructions.

Early in one's first space mission, the up/down ideas can be disconcerting. I actually flew into the ISS from the space shuttle *Atlantis* upside down! Not realizing how important orientation was, I made a body twist somewhere in the bowels of the shuttle's docking tunnel and ended up flying into ISS thinking, "Why the heck is everyone upside down?" Shoot, turns out I was!

So we typically will stay in a feet-to-the-floor, head-to-the-ceiling type of orientation as that helps us to adapt to our new environment. Early in a mission it can be disconcerting to some fliers when they are in a "normal" heads-up attitude and then do a quick flip to be upside down. Your brain takes a bit of time to recompute and understand what it is seeing, and it may lead to some "stomach awareness" and the urge to puke. But once you've been on board for a while, your brain begins to compute much faster, almost instantaneously. Then, all bets are off, and living, working, sleeping, and playing can be done in any attitude. That's the fun part!

Astronauts can work side by side in the same orientation or sometimes, in an effort to make the work easier, we can be totally opposite in our positioning. That's one of the beauties of zero-gravity: we can use the entire work space—that is, the entire volume of the modules—not just two dimensions as we do here on Earth.

If we are not allowed to touch or put our feet somewhere, those areas are clearly marked. They are actually rare on board and usually involve not blocking a port that discharges oxygen or is ready-made to accept the tip of a fire extinguisher in the event of an emergency.

Question: How do astronauts sleep while in space? Is there any sort of training for them to prepare for sleeping conditions?

Answer: The brief answer to your question? I slept fine! And, no, there is really no specific training that astronauts receive with regard to sleeping and sleeping conditions. As a matter of fact, the best advice I got for sleeping was from my STS-117 shuttle commander Sturckow. He told me to bring a good book and don't make any noise!

For those of you seeking more insight, my personal experience with sleeping in space was "dreamy." No, I didn't dream differently in space, but I did dream about stuff, just as I do here on Earth. As I discussed earlier, I was part of a special experiment called SLEEP (sleep-awake actigraphy and light exposure during spaceflight) Long and had to wear a special watch for the entire 152-day increment and many days before I flew and after I landed. The experiment and the watch measured light, darkness, and motion, providing investigators with key information about when I was sleeping and to what level. The watch's data could tell them when I was restless and when I was in a deep (REM) sleep. Petty cool. Postflight data revealed that I averaged about seven hours and twenty minutes of sleep over the mission, or much more (and better quality) than I think I am getting on Earth.

My sleep station on ISS, the temporary sleep station, is no longer there, but it was a precursor to the sleep stations now on board in Node 2 of the U.S. segment. The TeSS was quiet, dark, and cold. I liked it cold! We had Russian-made sleeping bags that just needed to be tied at four corners somewhere—and in any direction—to keep us from floating around during our sleep period. The bags were lightweight and just warm enough. I slept clad only in a pair of white Hanes boxer shorts. Many astronauts found it too cold on the ISS and slept in an assortment of long johns, pajamas, and even stocking caps.

36. Taking a snooze in the Quest air lock the night before embarking on another space walk.

In my nighttime routine I never used a computer in my sleep station. Evidently I was a forward thinker as pundits today tell us to keep our electronic devices out of our beds! I did all my work outside in the lab (teeth brushed, potty break taken, etc.), then turned off the lab lights before entering my "bedroom." After I climbed into my sleeping bag, with my head to the ceiling so the cold air-conditioning vent would blow my exhaled CO_2 away from my face—a key consideration for sleeping orientation—I placed foam earplugs in both ears and read a novel about a James Bond–like guy who chased historical artifacts until my eyelids got droopy.

Funny thing. When your eyes just start to close, your body relaxes and your hands release the book (well, mine did). But without gravity, your head doesn't fall and then jerk back up when you realize you've dozed off. If you do awaken, your book will still be floating there, right where you left it.

I never really took naps while on the ISS. If I felt myself starting to drift while working at the computer, for example, I would go get something to drink or eat in hopes of raising my energy level. I did not want to sleep during the day for fear of *not* being able to sleep at night. It seemed to be an effective strategy for me.

Question: What consumer products are actually used by astronauts in space?

Answer: Here's a short list of easily recognizable items that were with me on the ISS Expedition 15/16:

M&Ms in *all* forms
Jell-O Pudding cups
IBM ThinkPad (Lenovo) computers
Nikon, Kodak, and Canon camera equipment
Imodium A-D (thank God!)
Oreos and Chips Ahoy! cookies
Tabasco sauce
Cabela's pants and shorts
Land's End shirts
Victorinox Swiss Army knife
Omega Speedmaster X-33 watches
Hanes boxer shorts
ASICS running shoes (mine anyway)
Kodak film (yes, after all it was 2007)
Gillette Mach3 razors and blades
Colgate toothpaste
Velcro
plastic "zip" ties
iPads and iPhones
Huggies Baby Wipes
Tootsie Pops (one of my favorite snacks!)
Girl Scout cookies
Pentel mechanical pencils
Fisher Space Pens
Sharpies
3M Post-it Notes
Casio electronic keyboard

There were many others as well. And, yes, the shuttles did use Michelin tires.

Question: How much does an astronaut get paid during a mission on the ISS?

Answer: Not too much! Truthfully, we only get a small per diem for our time on the ISS. I believe my bank account had a deposit for about U.S.$172.00 after my 152 days in space. That comes out to about $1.13 a day. But when you consider that we have room and board provided, and we are not going out for dinner and drinks every night, it's probably a reasonable (but cheap) government estimate. Our meals, incidentals, and expenses don't amount to much while we are sailing around the Earth at 17,500 miles per hour. Overall, we receive our normal annual salary whether we fly in space or not, and—as you can clearly see—we don't get much extra when we do fly. It probably cost our government way more to calculate my per diem and have it electronically deposited into my account than the $172.00 I received!

I recall while on the ISS talking about this subject with my Russian crewmates. While not divulging too many details, I learned that their annual salary was about U.S.$6,000. For their six-month stint on board the ISS, they would make about ten times that salary.

Now you know a possible reason as to why the cosmonauts are so interested in flying multiple times. It can be a big advantage to their bottom line. In addition, if the cosmonauts performed a task—a manual Progress docking, for example—that wasn't on the originally developed time line or manifest, they would also receive additional funds. I speculated that another reason why they were very cordial with the ground controllers, no matter how upset they were in orbit, was that they didn't want to annoy anyone and jeopardize their pay. Too bad the United States didn't follow the Russian's lead. I might have kept my mouth shut a bit more.

Remember, that I flew on the ISS way back in 2007, so the numbers have probably grown higher.

Question: As an astronaut, when does it "hit" you that you are in space? At what time during the flight do you realize that you are actually in space for the first time?

Answer: I certainly can't speak for every astronaut, but I know what my experience was. Sitting on the launch pad inside the shuttle *Atlantis* (2007) and then *Discovery* (2010) for almost two and a half hours each launch, I had a lot of idle time. While listening to the somewhat mundane cadences in the background as engineers, technicians, and launch personnel assessed all systems and their readiness for launch, I spent that time trying to catch a few winks of sleep and anticipating what my initial moves would be upon reaching space for the first time. Relaxation was not in the cards you see, as my only thought then was not to screw anything up or to put our mission in jeopardy of any kind due to my not being ready to perform my roles and responsibilities.

But realizing I was in space for the first time was easy. There was *no way* I was not going to know, for I had the advantage of what we know as zero gravity.

As we rocketed into the heavens, I was seated on the shuttle middeck for both missions. Without any windows, my only connection to what was going on was my training knowledge and my understanding of the verbal "script" that I heard with each and every call to and from the ground and our commander. Through this nearly constant recitation of various milestones in our trajectory—"negative return, press to MECO [main engine cutoff]," and so on—I could tell how the shuttle was performing and where we were on our highly detailed and well-understood trajectory. But with the technical definition of "space" being 62 nautical miles (100 kilometers) above the launch pad, I knew we were getting closer and closer to the nirvana astronauts know as "orbit," but I had no real idea of when we would achieve that altitude defining us as astronauts unless I heard a callout from the commander or the ground. *They* had displays that provided the information. And I wouldn't feel the effects of zero-g until after the engines shut down. At that point, we were much higher than 62 nautical miles.

Question: Do astronauts ever accidentally leave items on the ISS when they come back to Earth?

Answer: What you hear isn't always correct. The ISS maintains a database called the Inventory Management System. Similar to an American grocery store, items are managed via computer and a scanner, using name, serial number, location, and so on to make sure that we know as best as we can where critical equipment and spares are located. Used items are scanned into "trash"; new items are scanned into their stowed locations. Moved items are unscanned from where they were located and rescanned to where they are going. That system, however, does not guarantee against losing things.

In a microgravity environment, things can get lost. As noted previously, while "showering" one morning I lost a chain with a St. Christopher's medal that Sunita Williams's mother had given to me before I launched. Turns out I found it floating near the floor behind one of our cargo stowage bags. Yay! Oftentimes the best place to look first is by an air vent where the airflow might have sucked your wandering item up against the grate of the vent. Air vents sometimes serve as our best "lost and found" location. Many things, as well as filth, turn up there.

Camera lenses and tools have been lost (both inside and outside of the ISS!). Many small items sometimes succumb to the lack of gravity and our "wind from the south" to send crews on wild goose chases as we try to locate them.

And even with this grand database—which, by the way, is comanaged by NASA and the Russian Space Agency—things can be overlooked or misplaced and larger items can be in the "failed to locate" pile for quite some time. While I served as a CAPCOM during Expedition 18, the crew called down looking for the electronic piano keyboard. No one (onboard *or* on the ground) could locate it via the Inventory Management System. I simply called up and said, "When I lived there, it was located in a big white bag, on the ceiling in Node 1, in the second rack section." One minute later, the crew called down saying, "Thanks, we found it!" So much for the concept of high-tech modernization.

Question: If you're acrophobic, how do you feel in space and in orbit?

Answer: First of all, I had to look up "acrophobic." Thanks for contributing to this retired astronaut's tightly bounded vocabulary. Once I learned that acrophobia is the fear of heights, I could attempt to answer your question. Are you ready?

My answer is, I don't know. As described in chapter 9, I am guessing I'm not acrophobic.

You'll have to find someone else to answer this question. My guess is that most (if not all) astronauts are not acrophobic, so it's gonna be hard to find an answer.

You might want to wait and pose the question to some of the tourists who are paying big bucks to fly with the Russians or one day with Virgin Galactic. I'm thinking they don't know either, but they will probably figure it out sometime prior to reentry.

Question: Is it okay to swallow toothpaste in outer space?

Answer: I found it *absolutely* okay to swallow toothpaste while living in outer space! I swallowed my off-the-shelf Colgate mint-flavored toothpaste at least 304 times during a 152-day stay in 2007. Downing your spit—laced with the minty-fresh flavoring from the toothpaste tube—may sound disgusting, but it actually can be considered as a space-age after-dinner mint.

The alternative of spitting multiple times into a towel or napkin is used by some astronauts who can't seem to abide the other option. I found that method to be quite messy and a waste of good consumables. Plus I didn't like having to manipulate a spit-filled towel in zero gravity. By the way, I have never heard of any special space toothpaste on the ISS that was specifically designed to be swallowed. During my flight, everything used was available in a grocery store on Earth.

It's now been about ten years since my first foray into the vacuum of outer space, and to paraphrase the great astronaut and America hero U.S. senator John Glenn, "Houston, I'm back home from zero-g, and I feel fine."

Question: Is being an astronaut the most technical job in the world?

Answer: I certainly don't think being an astronaut is the most technical job in the world. If it was, I am not nearly smart enough to do it. Since I had a successful fifteen-year career as an astronaut, with 167 days of living and working in outer space and taking space walks, there's no way in my mind that it can be considered the most techni-cal job in the world.

Most of you out there—with proper training, physical health, and attitude—would be excellent astronauts. And who knows? With all of the commercial spaceflight companies popping up, you may all get your chance.

Question: What is the best autobiography written by an astronaut?

Answer: Well, I'm certainly hoping the answer will be mine! While not an autobiography, my memoir, *The Ordinary Spaceman*, contains stories that are fun and thought-provoking, written in the same style as my answers here. So if you like what you are reading from me here, I am hopeful and confident that you will also like my first book. At the very least, give it some consideration.

Many other astronauts including the late John Glenn and Neil Arm-strong have written excellent books and memoirs. More recent books are from Astronauts Mike Massimino, Chris Hadfield, Scott Parazynski, Scott Kelly, and Leland Melvin. Mine's the best!

Question: What was it like to watch either the *Challenger* or *Columbia* space shuttle disasters live?

Answered: Unfortunately, I witnessed both of these disasters live—one via television and one in person. But I did so in very different contexts.

For *Challenger*, I was not yet an astronaut. In January 1986 I was still an aerospace engineer and working in the Mission Planning and Analysis Division at the Johnson Space Center. We were devising ways for space shuttles to approach and dock with some of the new and quite varied space station configurations: power tower, Delta, Space Operations Cen-ter, and others. I was seated in a meeting—discussing various aspects of how to safely approach and dock while using minimal fuel—on the sixth floor of Building 1, housing administration and management, on launch morning.

We took a break and turned our attention to the television set mounted near the ceiling in the conference room's forward right corner. As the events unfolded, nominally at first, it quickly became clear to me that something had gone wrong. But we all sat there in silence. No one spoke; no one moved. It was not until I heard screams and sobs emanating from the hallway that reality truly sunk in. My walk back to the office was somber. I felt helpless and alone. The remainder of my day was not good.

The *Columbia* accident took on a far more personal signature. I was one of the family escorts for the mission and had been handpicked by shuttle and mission commander Rick Husband.

Although I'm not sure exactly why he picked me, I was honored. It would be the first and *only* time I was ever asked to be a family escort. I believe it was because of my Christian faith—Rick was also a devout Christian leader—and my demonstrated successes while a not-yet-assigned astronaut that he gave me the nod.

As we fast-forward to landing day—February 1, 2003—I was hanging out at the crew family area, near the midpoint of runway 15/33 of the shuttle launch facility at the Kennedy Space Center. It was an absolutely gorgeous day for a shuttle landing. Steve Lindsey, the lead escort and a veteran space shuttle commander, and I were chatting with *Columbia* crewmember Laurel Clark's sister and enjoying the moment. As we watched the countdown clock intermittently and readied ourselves for what we thought was to be a wonderful landing and reunion, everything seemed to be progressing as planned. Of course, it didn't turn out that way.

A countdown clock that continued to increase past 00:00, a security guard's face that turned ghostly white in an instant, a lack of response from the *Columbia*'s crew to repeated calls from CAPCOM/Astronaut Charlie Hobaugh, and the words "get ready" uttered quietly by Lindsey—all told me that it was time to be strong in the face of true adversity.

It was the most difficult day of my life, even harder than when my father died, as that event was expected. I do my best to describe the day's events in *The Ordinary Spaceman*, but suffice it to say I felt a huge emptiness inside. It is a burden I still carry today. That day changed me; it made me much more emotional than I can ever remember being. It tested my faith, and it still tests me.

But it never dulled my dream to fly in space. It only strengthened my resolve. We at NASA would figure this out. We would make it safer than before. That's what we do. That's how we roll.

37. The crew of *Columbia*, STS-107.

Spaceflight is dangerous. Spaceflight is unforgiving of mistakes. We must keep that in mind as we move forward into the era of commercial spaceflight endeavors, lest we forget the lessons of our past.

Godspeed, *Challenger*. Godspeed, *Columbia*.

Question: Would it be possible for an astronaut to bring a paperback novel along on a mission?

Answer: Of course! As I noted in chapter 5, during my five-month stay on the ISS in June 2007, I read Clive Cussler's *Sahara*. It was my first, but definitely not my last, time with characters Dirk Pitt, Al Giordino, and the gang. I also read a copy of Carl Sagan's *Pale Blue Dot: A Vision of the Human Future in Space*. His book was a bit "meatier," but it was pretty cool to be reading his vision for humans in space while I *was* in space!

The ISS had a very small library of paperback books, but now almost everything has gone (as you might expect) electronic with e-books.

Question: What do astronauts do once they leave NASA?

Answer: Since I retired in January 2013, after a thirty-year career with NASA (fifteen as an aerospace engineer and fifteen as an astronaut), I can tell you exactly what *this* astronaut is doing since he left NASA.

Contrary to what many folks who know me might think, I am *not* sitting around shirtless in my recliner, wearing soiled boxer shorts, and scratching my hairy belly while downing a Miller Lite (although it might be fun to try that once or maybe twice). I have found that life after NASA and retirement are keeping me plenty busy, maybe busier than when I still worked *for* NASA.

Today I am employed part-time by Iowa State University where my title is distinguished faculty fellow, senior lecturer. In this capacity, I—along with two professors—team teach a freshman-level Introduction to Aerospace Engineering course, which includes aspects of spaceflight, airplanes, and generalities such as teamwork and ethics. I teach the spaceflight, teamwork, and ethics portions of the course and enjoy my time there. In 2013 we kicked off a prototype Spaceflight Operations Workshop (https://www.news.iastate.edu/video/view/id/bvmgvG1PghE) that was a huge success. It continues to be offered each summer and is growing in popularity and content.

In addition to my university duties, I became a first-time author in 2015 with the publication of *The Ordinary Spaceman: From Boyhood Dreams to Astronaut*. I am very excited about its success, as you may well imagine. I did *not* have a ghost writer, and I am extremely proud it is my work folks are reading. I have published a children's book as well. *A is for Astronaut: Blasting through the Alphabet* hit the shelves a few months prior to this one.

I also serve as a motivational speaker, voice-over persona, and consultant on many projects ranging from aerospace to biomedical engineering. I remain a staunch advocate for NASA and all it does, and I focus on leadership, teamwork, and STEAM (science, technology, engineering, art, and math) education. You may contact me through my website, http://www.astroclay.com/ or http://uniphigood.com/portfolio_page/clayton-astro-clay-anderson/.

Among my other projects I have also served as the global ambassador for the Giorgio Fedon 1919 Space Explorer brand of Italian watches. The watches, unveiled in Hong Kong for the first time in 2014, are available commercially.

Dang, it's time to get back to work.

Question: Can an astrophysicist go into outer space for a NASA mission?

Answer: Sure, as long as he or she has been selected as a U.S. astronaut. If you don't have that box checked, then you are out of luck for now. Many astronauts, including Doctors Stan Love and John Grunsfeld, have claimed astrophysics as part of their résumés and have performed quite well on space missions for NASA. So my advice to you, assuming you are not an astronaut but you *are* an astrophysicist, is to work at becoming an astronaut or—as my extremely intelligent friend Robert Frost suggests—earn millions of dollars and pay the Russians for a seat! For an astrophysicist on board the ISS, you might be bored as many of the windows face Earth.

Question: Are there any pictures taken from outer space in which you can see satellites orbiting Earth?

Answer: Well, I can think of a couple of examples.

The first would be any pictures taken of the Moon, which is one of our nearest satellites. But as I hear you mentally giving me a "humph," I would quickly turn to two *other* satellites that orbited Earth, both of which I was partially responsible for.

Dubbed Nebraska 1 and Nebraska 2 by ISS Expedition 15 commander Fyodor Yurchikin, the two satellites would orbit Earth for only a fraction of the time normally attributed to the more readily accepted American creations sent into orbit. Not launched into the geosynchronous realm of the Russian Global Navigation Satellite System (known as GLONASS) or the U.S. Tracking and Data Relay Satellites and Global Positioning System satellite, these low-earth orbit-launched, flying chunks of hardware were deemed no longer necessary to the life of the ISS. Officially called the Early Ammonia Servicer and the Vehicle Structure Support

Assembly (or something like that), both items were outdated parts of the ISS that I jettisoned during a seven-hour-and-forty-one-minute space walk on July 23, 2007.

The support assembly was the first to go. About the size of a large desk, it was the easiest of the two to chuck into orbit. Needing a velocity sufficiently high in the direction of Earth's center (radially) to ensure a de-orbiting scenario, I hurled it for all I was worth. The result was that a piece of space junk burned up upon entering Earth's atmosphere some three months later.

The servicer, weighing about fourteen hundred pounds, was not as straight forward a throw. It was the size of a double refrigerator or freezer, and its handrails offered me a solid grip. But those handrails were at nearly the full width of my space suit's reach capability. With my arms extended at forty-five-degree angles from my sides, I lacked the leverage and flexibility to give it a really good chest pass, as I used to do while playing basketball.

Flying 225 nautical miles above Earth on the end of the Canadarm2, my feet were locked solidly into the boot plate known as the articulating portable foot restraint. Positioned upside down—that is, with my head toward Earth—my job was to throw the servicer as hard as I could, again primarily toward the Earth's center. I had a mental forty-degree "cone"—with its vertex originating at my chest and its always expanding base—as my target. If I could throw it away fast enough and somewhere inside that cone, we would have success.

I rocked back in space, using a toe-heel motion I had practiced while wearing my space suit but lying on my side on the air bearing (think air hockey table) floor at the Johnson Space Center in Houston. The technique we designed was for me to lean back as far as possible and then, using my toe-heel move again, to initiate a forward motion. Exhibiting enough patience to allow me to go just slightly past vertical, I would heave the tank into the farthest reaches of the universe or at least away from me! As before, I had a target velocity, above which the control center team could breathe a sigh of relief. The servicer wouldn't recontact the ISS.

I was two-for-two that day, and my "pitching velocity" was far better than that of any major league baseball pitcher who ever lived.

38. Flying on the end of the Canadarm2, I move away from the ISS to jettison a piece of unneeded hardware.

Question: What's the best spot in Cape Canaveral to see the *Atlantis* launch?

Answer: Obviously, this question is a bit outdated, as *Atlantis* is now sitting silently but majestically in part of a marvelous display—one I would *highly* recommend you go see—at the Kennedy Space Center. Yet the question still has relevance today as launches of various vehicles still light up the skies of Cape Canaveral.

Launches are an amazing display of color, light, noise, and "made in America" technology. As a veteran of two liftoffs—the view's not as good inside, but the ride is spectacular!—and a viewer of numerous others, I say you should definitely add one to your growing bucket list.

Ideal viewing locations abound on the Florida coast. The best you could ever get was about three miles distant from the launch pad, but it still allowed one to see the beauty of the event and *feel* the tremendous pressure wave as it poured forth from the massive yet controlled explosions. The Kennedy Space Center has official viewing sites near the Saturn V facility that are available to anyone enjoying the Spaceport USA tourist experience. Contact Delaware North for more details and perhaps a launch schedule. And with all the beaches stretched along the Atlantic coastline, if your timing's good, then it's as simple as stopping your game of Frisbee for a moment to "catch one" on the fly.

Parks abound all around the area, including those near Titusville, Florida, and provide the opportunity for an almost football tailgate-like experience. There your entire entourage can enjoy the spectacle of a spaceflight launch and a manatee sighting while enjoying the trappings of an outdoor barbecue.

With the advent of a number of hard-charging commercial spaceflight companies—SpaceX, Boeing, Sierra Nevada, Orbital ATK—all seeking to continue U.S. leadership in space exploration, the coming years promise many awesome viewing opportunities from Florida to Wallops Island, Virginia, to Brownsville, Texas.

Question: What should I do to be an astronaut?

Answer: As noted previously, you should go to the NASA website (www .nasa .gov) and the USAJOBS website (www.usajobs.gov) and read all about the application process and the job requirements. Assuming you meet or will be able to meet the requirements, when a selection cycle is announced (check social media and the websites) and NASA needs more astronauts (not likely for a while), you should submit your application. Then sit back and wait.

Good luck. "May the Force be with you!"

Question: Is it mandatory for an astronaut to undergo two years of training?

Answer: If you are lucky enough to be chosen and NASA asks you to report to the Johnson Space Center in Houston as an astronaut candidate (AsCan), you have the opportunity to accept or decline. In my experience, rarely has anyone ever chosen the latter. So, once you arrive, you do all of the things they require of you training-wise.

The first two years for AsCans of civilian *or* military vintage are spent visiting NASA centers throughout the United States and learning more general skills, such as flying the T-38, Russian language, public affairs, basic ISS systems orientations, and so forth. If you are fortunate enough to survive that barrage (which also includes a survival-training stint), then you will graduate to the status of "unflown astronaut." For recent AsCan selections, with smaller class sizes, the basic training has been completed more quickly, or in less than two years.

At this point you are officially eligible for a flight assignment. While not very likely with today's long line for an ISS mission slot, you are instead given what astronauts affectionately refer to as your "desk job." These jobs provide support to astronauts already in space and to those in the specific training flow because they have been assigned to an actual mission. Your desk job may find you training to become a CAPCOM, the astronaut in Mission Control who speaks directly to the onboard crew, or working with advanced vehicles such as SpaceX, Orion, CST-100 Starliner, and Sierra Nevada. As a newbie, it is *highly* unlikely that you will be assigned to work in the areas of spacewalking or robotic arm operations but not impossible.

Your performance in these various jobs goes a long way in establishing your astronaut reputation among the various NASA communities at the JSC and potentially enhance your odds to be assigned to a mission. Piss people off? You will wait a long time. Do what you're told and do it with professionalism and a smile? An assignment could be forthcoming. It's all "black magic" to me, and lobbying—that is, politics—goes on. Trust me on that one.

So, in answering your specific question—and guessing that you are an AsHo (astronaut hopeful)—I would lose any arrogance in thinking that you already have most of the knowledge that astronauts do. That's not a good way to start your pursuit of *this* dream.

And, yes, astronauts—military or civilian—get paid for doing this training. Like anyone else in the job market, they are paid for their services, which—for most of their careers—include continuous training for, and maintenance of, their very specific skill set.

Question: What will happen if a person doesn't match the requirements to be an astronaut after being one but prior to testing again?

Answer: If I understand your question (and it is entirely possible that I don't), no astronaut needs to worry about matching requirements after having gone to space. For example, regarding one's height, *all* astronauts increase in height due to the absence of the full force of gravity while in space, but they return to their normal height almost immediately upon landing.

Aside from general medical maladies that could occur for whatever reasons—for example, kidney stones, heart murmurs, cancer—the only issue I know of that could ground an astronaut from returning to space after *already having been there* is radiation. If you return from a space flight (a long-duration ISS mission or, in the past, multiple shuttle short-duration missions) and your radiation exposure exceeds the documented, agreed-to limits, you will not be allowed to return to space. The radiation limit for females is different than for males (given reproduction issues), and that is the extent of my knowledge on the details of the subject.

I was launched early to the ISS in June 2007 with STS-117 (originally scheduled to launch in August 2007 with STS-118) to replace Sunita

Williams, as there was concern for her radiation exposure limits and that violating those limits might affect her opportunities to fly in space again.

Question: How does one go about becoming a mission specialist for NASA, and what experience and degrees are in demand and required for this position?

Answer: Step 1 is to go to NASA's website www.nasa.gov and look at the information provided there. You will have to fill out an astronaut application to have any hope of ever becoming a mission specialist. You must be a U.S. citizen, and your experience must be in the physical sciences. At a minimum a bachelor's degree will be required. To my knowledge, no specific degrees are in demand. Good luck!

Question: Do astronauts worry about their safety in space?

Answer: I never worried about my safety. As a matter of fact, once I arrived on the ISS with the STS-117 crew, the crew did a space walk to add an additional solar array cluster. Once the array was installed and unfurled, power began to flow. Well, that wasn't a good thing as the increased amperage burned out a box in the Russian segment—the box that controlled all six of its command and control computers! So, here we were, on the ISS all of four days, and all I could think of was, "I don't have to go home already, do I? I just got here!"

We usually don't worry about our safety. We trust the thousands of dedicated employees around the world who are doing their best to keep us alive. And they train the fear out of us! We are so worried about staying on our time line and following our procedures so as to not make a mistake that we focus on all of that and have no time to be afraid.

Question: Which time zone do the astronauts follow during their stay on the International Space Station?

Answer: The ISS follows Zulu time, or the same time as in Greenwich, England. That time zone, sort of in the middle of the major participating nations—the United States, Russia, Canada, Japan, and those of the European Space Agency—puts everyone except those at ESA at a small disadvantage. While a tiny price to pay for the countries involved, it makes for some sleepy folks at meetings! Most get used to the various time zone differences, such that business can be conducted appropriately. For Mission Control Center operations, all of the countries follow the Zulu time designations, with American flight controllers on three eight-hour shifts, 24/7, 365 days a year.

Interestingly enough, CAPCOMs (who are usually but not always astronauts) in the MCC who speak directly with the crew do not work the shift when the ISS crew is asleep. That is when most of the re-planning, if needed, takes place.

Question: What is the most severe injury to have ever happened to astronauts at ISS?

Answer: I don't know if you can qualify the inability to urinate as a "severe injury," but to my knowledge that is the most serious thing that I can think of having occurred in space. Upon reaching orbit, it is pretty critical that astronauts void their bladders as the experience of zero gravity causes your "I have to pee" sensors to behave not as reliably as they do on Earth. After having sat on the launch pad for a couple of hours, with your last shot at taking a leak being the bathroom at the 195-foot level of the launch tower, many astros are "ready to void" eight minutes and thirty-some seconds after launch. But their bodies may not tell them that.

In a particular instance that I know of, one of our astronauts required the use of a Foley catheter after reaching orbit. (As noted in chapter 5, we are trained to perform this procedure safely.) The insertion of this catheter let the astronaut empty the bladder and ultimately "get back to normal" a few days later. It did cause some hiccups on a scheduled space walk, so I would classify that as pretty serious.

Overall, the ISS is a pretty safe place, with scrapes, head bumps on handrails, and minor burns being the typical injuries. Gotta go. Have to go to the bathroom!

Question: Would NASA reject someone from the astronaut training program due to the spouse's occupation?

Answer: I can't say that I know the direct answer to that question, but I can speak from some personal experience about spousal occupations. When I was first selected as an AsCan in 1998, my wife worked for the Mission Operations Directorate in the Training Division. In their infinite wisdom, the Training Division's management and the Astronaut Office decreed that my wife was a threat to my training. Since the Astronaut Office had initiated protocols that we AsCans start taking exams to help qualify us as worthy candidates (there had been some problems with the selection of the previous class in 1996), the Training Division's management felt that given my wife's position in the organization, she could help me cheat on those exams and that she could get access to the training records and change my grades. Can you believe that? A pretty ridiculous circumstance if you ask me.

By the way, when I took my first exam as an AsCan—it was a general exam on the ISS—I was the first student to exit the testing room. I rose from my seat about twelve minutes after having first entered the room, with my test completed and after going over my answers three times. So what was my personal grade? 100 percent. I didn't need to cheat.

Question: What is the most foolproof way to become an astronaut?

Answer: I guess I'd have to say paying the Russians millions of dollars. If you have the bucks, odds are good that they have a seat for you. Choosing to go the conventional way of the American astronaut? Nothing foolproof about that. It takes hard work, good looks (thank you very much!), lots of luck, and help from others. Seriously? The best way is to check out the official requirements on the federal government's official jobs website. Submit your application when it's called for and take your chances. As noted earlier, the 2017 selection process received more than 18,000 applications for a meager twelve astronaut slots. So start saving your dollars! For additional tips, you can read about how I did it in my book *The Ordinary Spaceman*.

Good luck in your quest, and remember to keep lookin' up!

Question: What is the most difficult aspect of an astronaut's job, and why?

Answer: In my personal experience, there were many, as I'm really not *that* smart. But two very difficult aspects jump to the top of my list.

First was family separation. As a devoted husband and father, month-long trips to foreign countries such as Russia, Japan, Germany, and Canada were tough for me. Granted, they were very cool places to visit and train, but the routine of spending four to six weeks there, returning home for four weeks, and then doing it all over again for about three and a half years was hard. The other thing that was tough for someone with a small brain such as mine was the requirement to retain a tremendous amount of information on a tremendous amount of subjects: space walks, emergencies, Soyuz capsule, station systems, payload experiments, Russian language, and so forth. Fortunately, the MCC ground team was always available to help. That might not be the case for a trip to Mars.

Question: Who is the "arrogant and confrontational" astronaut that Chris Hadfield talks about in his book *An Astronaut's Guide to Life on Earth*?

Answer: I am not at all sure to whom Chris is referring, but there have been plenty of arrogant astronauts, male and female, in the corps over the years. Many have used the "don't you know who I am?" response in both work and personal situations, and that type of attitude is sometimes, but not always, denounced. Arrogant and confrontational astronauts have often been rewarded with additional spaceflights. I would guess that today, this is *not* the prevalent attitude, but it does still exist. Ego is a powerful thing if not controlled.

Question. What type of non-work pursuits would help set an astronaut applicant apart?

Answer: I always tell people to follow a simple plan. Find out what you love to do in life, with no expectations of ever becoming an astronaut. If you love to do something and you pursue that with passion, you will be one of the best at doing it. If you are one of the best, your chances of someday becoming an astronaut improve. If you want to learn how to SCUBA, fly a plane, or climb Mount Everest, do it because that is a dream you've always had and *not* because you think it will make you an astronaut.

Remember, I applied fifteen times to become an astronaut and was "kicked to the curb" by NASA fourteen times before being chosen. That year, 1998, twenty-five Americans were selected from more than 2,650 applicants. Do the math, and you find that's about 1 percent. There is no set formula to follow, but I wish you the best in your quest. I had a lot of luck, some good timing, and solid support from my family, friends, and community. But most important of all, I persevered.

Question: From which universities have the most astronauts graduated?

Answer: Statistics from the NASA *Astronaut Fact Book*, dated April 2013, show the following:

U.S. Naval Academy	52
U.S. Air Force Academy	36
Massachusetts Institute of Technology	34
U.S. Naval Post Graduate School	32
Stanford	21
Purdue	20
US Military Academy	18
Georgia Tech	14
University of Colorado	14
Air Force Institute of Technology	12
University of Texas	12
University of Washington	12

And, of course, Hastings College (of Nebraska) and Iowa State University have one, and that would be yours truly. We are currently working on increasing those numbers.

Question: What are the conditions for becoming an astronaut?

Answer: There are no conditions, just medical and educational thresholds to meet. For the clearest, most concise list of the requirements facing a wannabe astronaut, I offer two choices—buy my book or go to the federal government's official jobs site and search for "astronaut." The requirements will be clearly laid out there. Choice 2 will provide the most recent information. If you choose option 2, you can *still* buy my book and compare. Just sayin' . . .

Question: Six NASA astronauts are from the state of Ohio. What is it about that state that makes people want to leave the planet?

Answer: It's Ohio, for goodness sake! Perhaps just the act of living there makes people want to leave the planet. Suffering through the annual woes of the Cleveland Browns, Indians, and Cavaliers (before LeBron James) could be exactly what drives them away!

Dubbed by some as the "astronaut machine," Ohio is, and should be, very proud of its famous space travelers. From Armstrong to Glenn to Michael Foreman, Sunita Williams, and Donald Thomas, Ohioans have long been a source of America's finest space travelers, boasting twenty-five natives and thirty-four astronauts with Ohio ties. C'mon, Nebraska, let's get with the program!